Advances in 21st Century Human Settlements

Indexed by SCOPUS

This Series focuses on the entire spectrum of human settlements – from rural to urban, in different regions of the world, with questions such as: What factors cause and guide the process of change in human settlements from rural to urban in character, from hamlets and villages to towns, cities and megacities? Is this process different across time and space, how and why? Is there a future for rural life? Is it possible or not to have industrial development in rural settlements, and how? Why does 'urban shrinkage' occur? Are the rural areas urbanizing or is that urban areas are undergoing 'ruralisation' (in form of underserviced slums)? What are the challenges faced by 'mega urban regions', and how they can be/are being addressed? What drives economic dynamism in human settlements? Is the urban-based economic growth paradigm the only answer to the quest for sustainable development, or is there an urgent need to balance between economic growth on one hand and ecosystem restoration and conservation on the other – for the future sustainability of human habitats? How and what new technology is helping to achieve sustainable development in human settlements? What sort of changes in the current planning, management and governance of human settlements are needed to face the changing environment including the climate and increasing disaster risks? What is the uniqueness of the new 'socio-cultural spaces' that emerge in human settlements, and how they change over time? As rural settlements become urban, are the new 'urban spaces' resulting in the loss of rural life and 'socio-cultural spaces'? What is leading the preservation of rural 'socio-cultural spaces' within the urbanizing world, and how? What is the emerging nature of the rural-urban interface, and what factors influence it? What are the emerging perspectives that help understand the human-environment-culture complex through the study of human settlements and the related ecosystems, and how do they transform our understanding of cultural landscapes and 'waterscapes' in the 21st Century? What else is and/or likely to be new vis-à-vis human settlements – now and in the future? The Series, therefore, welcomes contributions with fresh cognitive perspectives to understand the new and emerging realities of the 21st Century human settlements. Such perspectives will include a multidisciplinary analysis, constituting of the demographic, spatio-economic, environmental, technological, and planning, management and governance lenses.

If you are interested in submitting a proposal for this series, please contact the Series Editor, or the Publishing Editor:

Bharat Dahiya (bharatdahiya@gmail.com) or

Loyola D'Silva (loyola.dsilva@springer.com)

Charles Chavunduka · Innocent Chirisa
Editors

New Urban Agenda in Zimbabwe

Built Environment Sciences and Practices

Springer

Editors
Charles Chavunduka
Department of Architecture and Real Estate
Faculty of Engineering and Built Environment
University of Zimbabwe
Mt. Pleasant, Harare, Zimbabwe

Innocent Chirisa
The Vice Chancellor's Office
Zimbabwe Ezekiel Guti University
Barrassie Road Off Shamva Road
Bindura, Zimbabwe

Department of Urban and Regional Planning
University of the Free State
Bloemfontein, South Africa

ISSN 2198-2546 ISSN 2198-2554 (electronic)
Advances in 21st Century Human Settlements
ISBN 978-981-97-3198-5 ISBN 978-981-97-3199-2 (eBook)
https://doi.org/10.1007/978-981-97-3199-2

This Springer imprint is published by the registered company Springer Nature Singapore Pte Ltd.
The registered company address is: 152 Beach Road, #21-01/04 Gateway East, Singapore 189721, Singapore

Editorial Board

Preface

The New Urban Agenda's potential contribution to the built environment sciences and practice in terms of skills, methods and processes has not been explored in the Zimbabwe context, yet towns and cities are engines for sustainable development. Attention to the built environment disciplines and practice is important because of their centrality to economic dynamism and sustainability of urban services. This collection, contributed from different built environment disciplines and professions, seeks to support the UN-Habitat New Urban Agenda passed at Habitat III in Quito, Ecuador, in 2016. The book begins by exploring theoretical and conceptual material on the built environment concept. It recognizes that sub-Saharan African countries and Zimbabwe in particular are operating in a changing environment of rapid urbanization, unemployment, poverty and inequality. Its premise is that the built environment sciences and practice can make a contribution to addressing sustainable development challenges through the introduction of technology and innovations in infrastructure, institutions and processes that enhance socio-economic growth and development. The book brings together scholarship from various disciplines in the built environment and presents sector initiatives such as those relating to curricula and new practical approaches for classroom application, the application of emerging technologies like blockchain, smart contracts, artificial intelligence and the use of remote sensing. It offers a renewed built environment concept, which is interdisciplinary in approach, one that redefines the discipline and practice pillars and their economic, social, environmental and governance dimensions for sustainable development. Hitherto, in the Zimbabwe context, approaches in built environment teaching and professions have lacked an integrated view that leverages co-innovation in the development of sustainable cities and communities.

The book is aimed at readers from disciplines in spatial planning, land surveying, architecture, real estate, geography, construction and civil engineering, quantity surveying, related sciences; and practitioners, including those working within international organizations, NGOs and built environment consultants. It fills a gap by connecting the hitherto sparse and sometimes fragmented built environment sciences and practice with the sustainable development agenda. In that way, it strengthens understanding of the New Urban Agenda and Sustainable Development Goals in

Zimbabwe. New Urban Agenda issues are discussed in the context of the actual planning, design and implementation of the built environment, highlighting both positive and negative practices. It further aims at purposefully influencing the New Urban Agenda and implementation of Sustainable Development Goals for success. With a focus on Zimbabwe, it highlights the opportunities and challenges faced in creating the built environment; and addresses how built environment sciences and practice can be shaped to steer the New Urban Agenda. Simultaneously, it analyses the reverse relationship, as through the provision of international norms, aspirations and guidelines; the New Urban Agenda positively influences the built environment disciplines and practice. It advances scientific discourse on the built environment concept and its effective application to the realization of the New Urban Agenda in Zimbabwe. Among novel contributions, the book offers a built environment climate resilience framework with a comprehensive platform that is accessible to developers and decision-makers. It identifies initiatives including those that harness smart technologies, academia thrust on innovation and business linkages, as well as those that have looked into shared value creation in the built environment.

Harare, Zimbabwe
Bindura, Zimbabwe
January 2024

Charles Chavunduka
Innocent Chirisa

Contents

The Built Environment in the Context of the New Urban Agenda: An Overview

Charles Chavunduka, Innocent Chirisa, and Tsungirayi Diana Tsikira

Abstract Zimbabwe is experiencing rapid urbanisation that is potentially unsustainable given the economic, social, environmental and governance context. The built environment sciences and practice can make a contribution to addressing sustainable development challenges through the introduction of innovations in infrastructure, institutions and processes that enhance economic growth and development. The book is aimed at readers from disciplines in spatial planning, architecture, real estate, geography, construction and civil engineering, land surveying, quantity surveying, related sciences, professionals and practitioners. The built environment has received increased prominence because of its potential in steering the New Urban Agenda through emerging discourses, institutions and practices; yet in most of sub-Saharan Africa and Zimbabwe in particular, the concept has not been fully developed. This book brings together scholarship from different disciplines in the built environment and presents sector initiatives such as those relating to curricula, smart technologies and cutting-edge innovations in steering sustainable development. It presents initiatives in various sectors of the built environment that have the potential to boost the competitiveness of cities and ideas on how the disciplines can be shaped for steering the sustainable development agenda. The chapter contributions in this book are outlined and how they contribute to evolving ideas on the built environment.

Keywords Built environment · New urban agenda · Curriculum · Practice · Zimbabwe

---

C. Chavunduka (✉)
Faculty of Engineering and the Built Environment, University of Zimbabwe, Mt. Pleasant, Box MP 167, Harare, Zimbabwe
e-mail: cchavunduka@eng.uz.ac.zw

I. Chirisa
Zimbabwe Ezekiel Guti University, Bindura, Zimbabwe

Department of Urban and Regional Planning, University of the Free State, Bloemfontein, South Africa

T. D. Tsikira
Department of Architecture and Real Estate, University of Zimbabwe, Mt. Pleasant, Box MP 167, Harare, Zimbabwe

C. Chavunduka and I. Chirisa (eds.), *New Urban Agenda in Zimbabwe*,
Advances in 21st Century Human Settlements,
https://doi.org/10.1007/978-981-97-3199-2_1

1 Introduction

The idea for this book has its origin in 2019 collegial discussions about the need to establish a faculty of the built environment at the University of Zimbabwe. All built environment disciplines would be offered in one faculty thereby improving the arrangement hitherto whereby some built environment disciplines like spatial planning and real estate had been offered in the faculty of social sciences. At that time, a countrywide universities curriculum review was undertaken with the objective of having the education system contribute to the transformation of Zimbabwe into an upper-middle income economy by 2030. The curriculum was to respond to the national industrialisation and modernisation agenda. In that respect, education should to go beyond the theoretical epistemic where knowing is a dead end in itself, to transforming knowledge into 'goods and services' through innovation for industrialisation.

During the national curriculum review, engagement in the built environment cluster raised concern with the fact that disciplines have been operating in isolation. In view of this limitation, the curriculum development and review process debated the adoption of inter and trans-disciplinarity in the built environment concept. This entails a new trajectory, moving away from discipline-specificity that excludes modules from other relevant disciplines to include such as long as they enhance the practical application of the degree qualification in real life. For example, a degree in civil engineering should also include modules in spatial planning and accounting. Thus, debate focused on the use and relevance of the interdisciplinary approach in the built environment concept.

Debate about the built environment disciplines coincided with related discussion concerning the matter that they fall under different professional institutions. The arrangement has not enabled interested built environment disciplines to participate in activities of common concern. An example is the ongoing revision of the 1977 Model Building By-Laws by engineers, a matter that is of interest to planners and architects. The lack of a coordination mechanism among professional institutions has precluded essential collaboration that would enhance built environment practice.

Whilst reviewing the built environment curriculum it became clear that it needs to leverage sustainable urban development, hence the New Urban Agenda (NUA). Essentially, SDG 11 aligns with the NUA but Zimbabwe policy documents are silent about the agenda. The National Development Strategy 1 (2021–2025) makes reference to inclusivity and resilience building but does not address other aspects of SDG 11. This book identifies gaps in the New Urban Agenda in Zimbabwe and explores how various built environment disciplines can be shaped to steer the process. It is aimed at readers from disciplines in spatial planning, architecture, real estate, geography, construction and civil engineering, land surveying, quantity surveying, related sciences, professionals and practitioners.

2 Legacies in Built Environment Education

Although Zimbabwe withdrew from membership of the Commonwealth in 2003, as a former British colony, built environment knowledge and practice were modelled on British higher education (Home 2013). There is a Eurocentric bias in knowledge production in the built environment disciplines that is a legacy of colonial building traditions. In fact, universities in anglophone Africa are strongly influenced by models inherited from the colonial era. Former British colonies have been members of various Commonwealth Associations of civil engineers, architects, land surveyors and planners and the professional training model has been based on British practice. For example, the Association of Surveying and Land Economy (CASLE) was formed in 1969 as a federation of independent professional societies representing surveying and land economy in Commonwealth countries. It develops professional bodies in its field and sets standards for all specialisms in surveying. The Commonwealth Association of Planners was formed in 1972 with the objective of developing skills of urban and regional planners across the Commonwealth to meet the challenges of urbanisation and sustainable development of human settlements.

Through the influence of Commonwealth institutions such as the Royal Town Planning Institute (RTPI), the built environment curricula in Zimbabwean universities have remained similar to those taught across the Commonwealth. Before the Government of Zimbabwe established planning schools, most practising planners sought membership of the RTPI in order to strengthen participation in Commonwealth built environment professional networks. The legacy of Commonwealth Associations may inhibit progress in evolving and advancing knowledge suitable for the Zimbabwe context of rapid urbanisation and rising levels of poverty in urban and rural areas. Apart from the influence of Commonwealth institutions, the legacy of colonial laws has been slowing the adaptation of built environment curricula to interdisciplinarity. Built environment curricula in African universities and Zimbabwe in particular have been slow to adapt and evolve, but curriculum development is beginning to adopt interdisciplinary approaches, and the breaking down of disciplinary boundaries to solve common problems (Home 2021).

3 The New Urban Agenda and Sustainable Development Goals

The New Urban Agenda (NUA) presented at the UN-Habitat III Conference held in Quito, Ecuador, in October 2016, is mainly aligned with Goal 11 of the United Nations Sustainable Development Goals (SDGs), which is to turn urbanisation into an engine for development and to make cities more inclusive, safe, resilient and sustainable (Caprotti et al. 2017; UN-Habitat 2016). It focuses on three aspects, which are to promote sustainability and inclusion in society to end poverty, promote inclusive and sustainable urban wealth and promote resilience and sustainability in development and the environment. It is a framework for guiding countries and their urban areas towards sustainable urbanisation. It is not legally binding and does not have an implementation plan. As such, different countries have been tackling the NUA in their own ways.

The 2030 global agenda for sustainable development stresses the importance of rapid urbanisation as both a challenge and solution to the most pressing global issues (Robin et al. 2019). Though cities only occupy 2% of the total land mass, they generate 70% of global GDP, consume over 60% of global energy, emit 70% of greenhouse gases and generate 70% of global waste (IISD 2018). Seeing as cities can generate wealth and grow the economy, the idea behind the NUA is to turn urbanisation and cities into tools for sustainable development, which in turn leads to job creation, poverty reduction and increased quality of life. There are too many people living in poor-quality housing and without adequate infrastructure services such as water, sanitation and electricity, without stable employment, reliable sources of income, social services or prospects for upward social mobility (UN-Habitat 2016). Informal settlements are home to about 62% of African urban population (UN-Habitat 2013) and informality constitutes the dominant form of contemporary urban living (Davis 2006). The NUA represents a paradigm shift in several ways. Firstly, it is a shift towards a new model of urbanisation that can better respond to the challenges of the twenty-first century, that is, issues such as inequality, climate change, informality in the urbanisation process and in job creation, and the unsustainable forms of urban expansion. Secondly, it is a shift away from the rigidity of the technocratic generic modernist model towards a more open, malleable and incremental urbanism that recognises the role of space and place and how they are shaped by planning and design—in making cities more equitable. Thirdly, it represents a paradigm shift away from the specialist-led approach to urban development embodied in the Charter of Athens towards a much more participatory process of urban co-creation (Mehaffy and Haas 2020).

The basis for the NUA is the many SDGs that can be met with sustainable urban policies and good local governance (Satterthwaite 2016). National urban policy is a key component of the NUA and is defined as policy specifically created to directly affect the development of urban areas (van den Berg et al. 2007). The other key components are rules and regulations, territorial planning and design, and municipal finance. National urban policies meet national development targets with legislation playing an important role to achieve orderly development. The NUA puts emphasis on regulated spatial planning, along with concepts of territorial and regional balance. Ultimately, the NUA projects international commitment to action on urbanisation as a global concern. It is a vehicle for achieving the SDGs and in particular Goal 11. Cities have a central role in the achievement of SDGs and if the NUA were to focus narrowly on SDG 11, a holistic approach to urban development would be lost (Caprotti et al. 2017). There is complex interaction among the SDGs which should be taken into consideration for effective implementation of the NUA. Underpinning the NUA is the assumption of the existence of strong, well-resourced and well-capacitated governments at national, regional and local levels; able to develop and implement policies and plans, and drive forward what is largely a state-directed agenda (Watson 2016).

4 New Approaches to Teaching

The concept of the built environment appears to have been popularised in the 1960s, but a review of the literature shows that developing a shared understanding of it remains a challenging task (Moffatt and Kohler 2008). The conceptualisation of the built environment has changed over the decades (Srinivasan et al. 2003). In its initial definitions, the built environment focused on buildings, their planning, creation and occupation. This approach focused particularly on practicality and measurable performance (Davis et al. 1993; Lutzkendorf and Speer 2005). During this period, the idea of the built environment was mainly limited to buildings and other forms of infrastructure such as roads, bridges and dams. This view dates back to the city beautiful movement (Wilson 1994) whereby it was believed that making American cities aesthetically pleasing would encourage the wealthy to spend their money there (Tougwa 2015). It was a materialistic approach, one which made it easier for policies to be changed according to the demands of the market or economy at any given time. Little thought was given to whether the physical infrastructure that was being added would actually benefit the broader urban society.

With time the idea of the built environment as a physical artefact shifted to being a process that is both directly influenced, and indeed created by actors, as well as the broader socio-political-economic climate (Nielsen and Farrelly 2019). It is both malleable and flexible to change through the agency of professionals such as spatial planners, urban designers and architects, whilst also being fixed and obdurate once physically built as a form of socio-technical lock-in. It is produced by society, whilst simultaneously influencing the actions and decisions of society. The built environment can reinforce society's sense of place, convey power and dictate what is acceptable and what isn't. It can be used as marketing, enticing visitors and new residents. It can influence user decisions and behaviour, that is, more bikeways, dense matrix of individual shops, walkways encourage cycling and walking versus driving.

Recent conceptualisation of the built environment views it as a system. It is a complex and interconnected system made up of buildings, people and services which consume, generate and import resources to create wealth, as well as dispose of the waste created during the process (Moffatt and Kohler 2008; Nielsen and Farrelly 2019). This expansive notion of the built environment conveys a systems approach, where dynamic relationships exist between built environment elements (Moffatt and Kohler 2008). Here the built environment is conceptualised as a socio-ecological system. It is constantly changing, reflecting the evolution of social systems and influencing this evolution in turn. Adopting a systems perspective to the built environment defines it not as an object but as social-ecological system. The built environment is a system and has material flows that begin with inputs from nature, then flow into intermediary processes. After use, flows may be reconverted by infrastructure systems for reuse. Ultimately, all flows are directed to waste products. Economic mainstream theory presents relations between the built environment and the ecosphere as exchanges between two equivalent systems. The built environment only exists in relation to the 'non-built' environment, that is, the ecosphere. It represents physical artefacts that embody a diverse array of overlapping systems at a point in time that have created the final outcome.

Contemporary thinking about built environment sciences and practice has seen persuasion in the Biglan disciplinary model which defines the built environment as an applied, but theoretically coherent, interdiscipline with a common epistemological axiomatic (Chynoweth 2009). The Biglan model develops its ideas from Kuhn's (1962) concept of a paradigm. Science proceeds, not through a process of incremental development, but by periods of uneventful 'normal science' interspaced by periods of rapid change (or paradigm shifts) following a crisis in the prevailing epistemological and methodological paradigm. Within this thesis, it is noted that different academic disciplines are characterised, to varying degrees, by the presence of paradigms that prescribe the appropriate problems of study and the validity of the methodologies to be employed. Some fields (typically the natural sciences) are characterised by highly developed paradigms, but others (e.g. the humanities) are less so.

The Biglan model draws on the concept of a paradigm which it uses to place disciplines on a continuum from 'hard' (paradigmatic) to 'soft' (non-paradigmatic). The model also explicitly incorporates the pure/applied knowledge theme which enables it to identify any discipline on a hard-soft/pure-applied matrix. Based on Biglan's empirical findings, the position of individual academic disciplines can be plotted on a matrix as illustrated in Fig. 1.

In Fig. 1, the natural sciences fall into the bottom left (hard-pure) quadrant whilst the arts and humanities are found in the bottom right (soft-pure) part of the matrix. In fact, there is a continuum from the natural sciences on the far-left-hand side of the diagram, through the social sciences in the centre, to the humanities and finally to the arts on the far-right-hand side. 'This transition reflects the progressive relaxation of paradigmatic requirements and the increasing level of personal input by the individual scholar into the academic enterprise' (Chynoweth 2009; 304). The matrix is completed by the inclusion of the applied axis. In this conceptualisation, the applied sciences, serve the engineering professions and appear in the top left (hard-applied) sector whilst the social and creative professions are found in the top right (soft-applied) quadrant.

In Zimbabwe, the built environment subject area is not well-established as a recognised field of study. Zimbabwe universities define the field as including engineering, architecture, urban and regional planning, quantity surveying, land surveying and real estate. This definition describes the field in terms of its various fields of application, rather than defining its cognitive base. Most undergraduate degree programmes in these disciplines operate in isolated silos (Ilesanmi 2017). They are mainly based on discrete curricula which prepare students as professionals: engineers, architects, planners, quantity surveyors, realtors and land surveyors—with limited interdependencies. The curriculum content of built environment undergraduate programmes

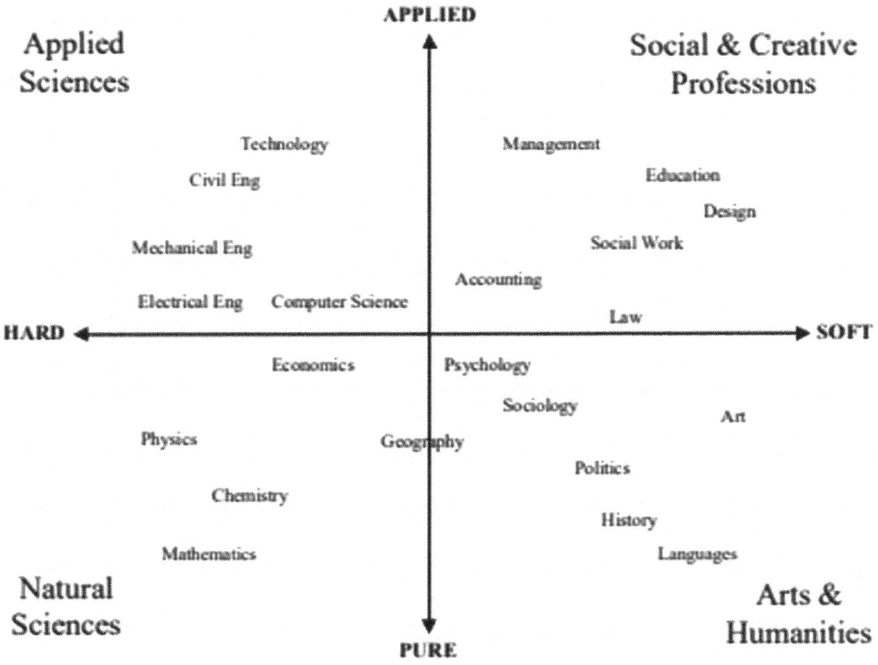

Fig. 1 Biglan disciplinary model. *Source* Chynoweth (2009)

provides an indication of the relevant areas of expertise. Various professional institutes have defined its academic base by reference to particular areas of knowledge. Although there are inevitably minor differences in the various descriptions, a degree of consensus is seen to emerge regarding the substantive areas of built environment knowledge. For the purpose of this book, these are defined in terms of the following six subject disciplines: construction and civil engineering, real estate and land economics, architecture, spatial planning, quantity surveying, geoinformation and land surveying.

The predominantly applied nature of the field's knowledge base can be illustrated by locating these areas of knowledge within the Biglan model (Fig. 2).

This exercise also highlights the enormous diversity of academic practices within the built environment which are seen to span almost the entire spectrum of the arts and sciences. This latter point raises questions as to whether it is appropriate to describe the field as an academic discipline at all, or whether it is simply an amalgamation of disciplines which collectively serve the fields of application identified above.

The built environment field is not a discipline in the sense of having an epistemic community sharing a unified knowledge domain. Also, the field is too diverse to be described as an academic discipline in its own right. Yet problems of sustainable urbanisation transcend conventional disciplinary boundaries and require an interdisciplinary approach. Initiatives to improve the urban environment and the poor need

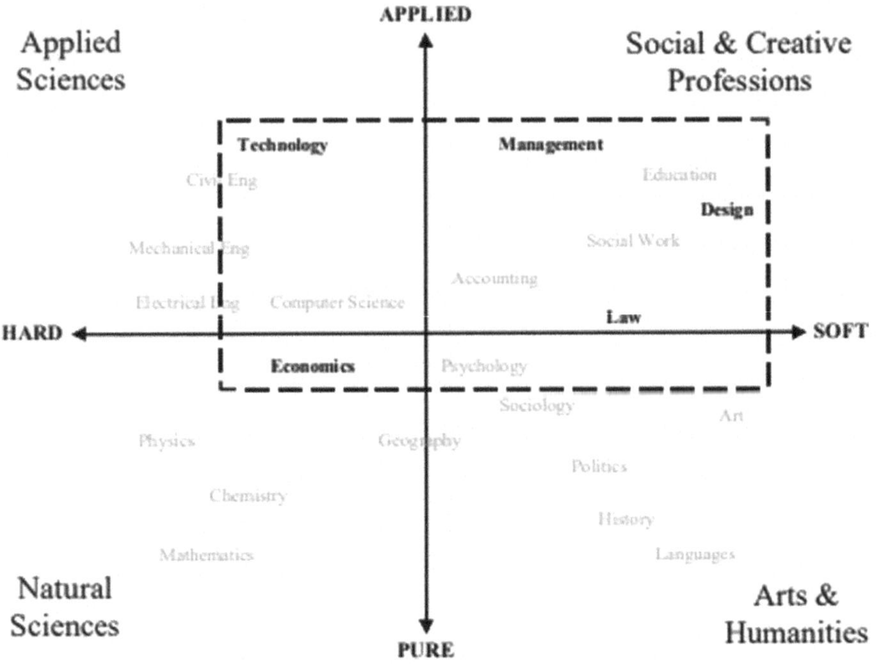

Fig. 2 Built environment knowledge base. *Source* Chynoweth (2009)

a holistic approach, more so, in 'overheating' Southern cities that are 'deindustrialising' and where populations are increasing (Evans and Marvin 2006). These complex societal problems do not respect disciplinary boundaries and are better understood and solved through an interdisciplinary approach (Stock and Burton 2011). Thus, interdisciplinary knowledge, skills and education are essential to meeting the shortage in high and middle-level human resources needed for the NUA.

There is an evident quest for interdisciplinarity but the Zimbabwe built environment is characterised by multidisciplinarity. Multidisciplinarity occurs where a variety of disciplines are encountered simultaneously in circumstances where the possible relationships between them are not made explicit. This is frequently associated with undergraduate modules where much common ground exists among the built environment disciplines but there has been a very slow convergence in approaches. There are evident differences in paradigms, methods and approaches between engineering and other built environment disciplines which pose a challenge to interdisciplinarity. The issue of disciplinary loyalty is strong but breaking down of barriers is more urgent than ever despite difficulties.

Interdisciplinarity occurs where a number of separate disciplines surrender their own axiomatics and collectively define themselves by reference to a common strategic axiomatic. This takes place where the traditional disciplines of knowledge are brought together in structures which reflect 'basic themes of society or need areas' rather than their own disciplinary identities. The existence of a common axiomatic then facilitates epistemological integration as the disciplines collectively address the resolution of common problems. Where this occurs a new hybrid form of knowledge is created which is usually referred to as an interdiscipline (Klein 1990).

In Chap. 2, Muneta and Chirisa do a global analysis of the contribution of value chain process in the built environment. They reckon that building processes are very complex and hence need more interaction and collaboration between actors in the industry. Various opportunities in the built environment are identified for eliminating process and market barriers in organisational innovation leading to new business models focusing on collaboration between the different building actors and/or new construction value-chain-related ecosystems. Improved chain relations and overall chain performance are expected to yield tangible benefits in terms of economic performance and, in some cases, poverty reduction. A really sustainable value chain demands a profound understanding and insight into resource production and heritage, and a comprehensive and sound assessment of good governance, social consequences, working conditions and environmental management. They identify an important role of government as a major investor in the built environment and environmental value chain. Improvement in digital technologies is seen to play an important role in the development of intelligent business models and their use in understanding share value creation in built environment context and the well-being of society, users and residents.

In Chap. 3, Chamunorwa reminds us that sustainable urbanisation requires an active role of the biophysical environment in shaping the built environment. The chapter examines the biophysical environment and built environment nexus and finds opportunities for planning, designing and managing the latter using the former's components as the guiding framework. It makes recommendations for government to craft policies that integrate biophysical conditions of urban areas in planning, shaping and managing the built environment. Counsel is given on the need for planning frameworks to include issues that go beyond traditional site analysis.

In Chap. 4, Masarira introduces built environment sciences by exploring the role of geoinformatics and land surveying in the realisation of the NUA. The chapter highlights challenges arising from rapid urbanisation and informalisation, particularly economic inequality, high youth unemployment and environmental degradation. The cadastre is posited as an engine providing a base for land use planning, tenure security and upgrading of informal settlements. Cadastral information, topographic mapping and existing plans provide a database shared by surveyors, engineers, architects, planners, realtors and other professionals in the built environment. Innovations including the adoption of inclusive tenure approaches, flexible tenure models, fit-for-purpose registration and use of drone technology in informal settlement upgrading will leverage the ability of geoinformatics and land surveying in steering the NUA in Zimbabwe.

In Chap. 5, Chavunduka points out that spatial planning provides a framework for sustained and inclusive economic growth, realisation of adequate standards of living and working conditions for all segments of society and management of the built environment of cities and territories. The chapter explores initiatives in spatial planning particularly informal settlement upgrading that have been advancing the goal of achieving sustainable cities and communities. It further analyses how the initiatives have been constrained by lack of an effective strategy for implementing the NUA, outdated legal and institutional frameworks, and lack of regulation of the planning profession. A coherent and up-to-date spatial planning policy, legal and institutional framework including regulation of the planning profession are recommended for the NUA.

In Chap. 6, Shumba examines the contribution of construction and civil engineering to the NUA in Zimbabwe. The agenda is seen from the perspective of developing sustainable smart cities. Discussion of the contribution is done within the context of the National Development Strategy 1 and Zimbabwe Vision 2030. A series of case studies on construction projects, integrated water management, water and sanitation, and smart city initiatives are used to demonstrate progress towards the achievement of Goal 11 (sustainable cities and communities), Goal 6 (clean water and sanitation), Goal 9 (Industry, innovation and infrastructure) and other SDGs. The chapter underscores the need for an implementation plan for the success of the NUA in Zimbabwe and sub-Saharan Africa.

In Chap. 7, Zavirima and Mavhima consider the contribution of architecture to the NUA from sustainable development and smart cities perspective. The centrality of architecture in the built environment is explored through the architectural concept of housing. Architectural role in steering the NUA is constrained by self-interested institutions, outdated laws, regulations and by-laws that do not enable intra and interdisciplinarity, public participation, evolving digital and building technologies including building information modelling. The chapter recommends a review of the legal framework to enable public and professional participation, and adoption of affordable new technologies for delivery of housing infrastructure, smart buildings, sustainable cities and communities in the NUA.

In Chap. 8, Kanonhuhwa and Mashingaidze analyse rapid urbanisation, infrastructure decay and challenges posed by the increase in informal public transport in relation to improving access to affordable, secure and reliable public transport. Official policy has been to capacitate a wholly government-owned public transport operator but the operator has not coped with demand. On the basis of case studies of Harare and Mutare, the authors proffer policy advice that is centred on improving the efficiency of the urban transport system as a whole, shifting towards environmentally friendly modes of transport and fuel efficiency. This would be enabled through the adoption of instruments on land use planning, regulation of norms and standards and investment in a functional public transport system.

In Chap. 9, Moyo and Chigara address the role of quantity surveying education through key informant interviews with quantity surveying academia, professional body representatives, consultants and government officials. The contribution of quantity surveying to the NUA faces among others, challenges arising from limited vocational education and training in sustainable development, nascent profession, lack of coordination mechanism among built environment professional bodies and absence of an integrated Construction Act. This has seen quantity surveying lagging behind other built environment professions in steering the NUA. Perceptions shared by key informants call for strengthening education for sustainability as well as a responsive construction policy.

In Chap. 10, Kwangwama examines the role of real estate and land economics in driving the NUA in Zimbabwe. The chapter traces the history of real estate in Zimbabwe to inform the opportunities and constraints faced in implementing the NUA. The shift from segregation of land uses towards mixed-use development has been an important paradigm in real estate development bringing more inclusive, equitable and sustainable outcomes. Land is found to be a constraint to real estate development because unlike in the first two decades of independence, post-2000 fast-track land reform has allocated acquired peri-urban farms to housing cooperatives and land barons who have not improved it through land development services thereby compromising the delivery of inclusive, equitable and sustainable cities. The continued decline of the national economy negatively affected the availability of finance, new investment and a sustainable built environment. A stable macroeconomic environment and improved land governance are identified as essential ingredients for real estate to enable the NUA in Zimbabwe.

In Chap. 11, Mazongonda and Chigudu address outdated legal and institutional frameworks including planning models and prescribe revisions for realisation of the NUA. The existing framework is rigid and development control biased but given rapid urbanisation and informalisation trend in Zimbabwe, planning ought to expend resources on visioning the future African city rather than evictions, demolitions and removals. Drawing from country trends in informality, urban economics, technological factors, environmental conservation, planning process and governance quality; the authors analyse the forces at play in pursuing the NUA. Governance reform is recommended for sustainable cities.

In Chap. 12, Chavunduka turns to a discussion of climate resilience amidst the impact of extreme weather events and the role that built environment disciplines and practices can play in climate change mitigation and adaptation. The chapter highlights that it is imperative to factor climate resilience in planning as a means to attaining the global concern on sustainable development and the NUA. It recommends a climate resilience framework that offers a comprehensive platform for developers and decision-makers.

In the last chapter, the authors review some recent scholarship and stakeholder activity and speculate on the future of the African built environment over the remaining 8 years of the New Urban Agenda and beyond.

5 Conclusion

The chapter introduced issues surrounding the potential adoption of the interdisciplinary approach in the built environment. It found that colonial legacy in built environment education is a constraint to evolving interdisciplinary approaches. In Zimbabwe, built environment disciplines such as architecture, urban and regional planning, quantity surveying, land surveying and real estate operate in silos yet curriculum development is pursuing interdisciplinary approaches. A case is made for interdisciplinary knowledge, skills and education in leveraging the role of the built environment in steering the NUA.

References

Caprotti F, Cowley R, Datta A, Broto VC, Gao E, Georgeson L, Herrick C, Odendaal N, Joss S (2017) The New Urban agenda: key opportunities and challenges for policy and practice. Urban Res Practice 10(3):367–378

Chynoweth P (2009) The built environment interdiscipline: a theoretical model for decision makers in research and teaching. Struct Surv 27(4):301–310

Davis M (2006) Planet of Slums. Verso, London

Davis R, Shrobe H, Szolovits P (1993) What is a knowledge representation? A1 Mag 14(1):17–33

Evans R, Marvin S (2006) Researching the sustainable city: three modes of interdisplinarity. Environ Plan A 38:1009–1028

Home R (2013) Of planting and planning: the making of British colonial towns. Spon, London

Home R (ed) (2021) Land issues for urban governance in Sub-Saharan Africa (Vol. 2). Springer, Geneva, Switzerland

IISD (2018) UN grapples with intensified urban challenges, oversight needs for UN-Habitat. http://sdg.iisd.org/news/un-grapples-with-intensified-urban-challenges-oversight-needs-for-un-habitat/ Accessed 12 Dec 2021

Ilesanmi AO (2017) Sustainability, transdisplinarity and built environment research. Int J Dev Sci 6(2):66–69

Klein JT (1990) Interdisciplinarity: history, theory and practice, Detroit MI. Wayne State University Press

Kuhn T (1962) The structure of scientific revolutions. University of Chicago Press, Chicago IL

Lutzkendorf T, Speer T (2005) Alleviating asymmetric information in property markets: building performance and product quality as signals for consumers. Build Res Inf 33(2):182–195

Mehaffy M, Haas T (2020) New urbanism in the new urban agenda: threads of an unfinished reformation. Urban Plann 5(4):441–452

Moffatt S, Kohler N (2008) Conceptualising the built environment as a social-ecological system. Build Res Inf 36(3):248–268

Nielsen J, Farrelly MA (2019) Conceptualising the built environment to inform sustainable urban transitions. Environ Innov Soc Trans 33:231–248

Robin E, Steenmans K, Acuto M (2019) Harnessing inclusive urban knowledge for the implementation of the New Urban Agenda. Urban Res Practice 12(2):137–155

Satterthwaite D (2016) Editorial: a new urban agenda? Environ Urbanisation 28(1):3–12

Stock P, Burton RJF (2011) Defining terms for integrated (multi-inter-trans-disciplinary) sustainability research. Sustainability 3:1090–1113

Srinivasan S, O'Fallon LR, Dearry A (2003) Creating healthy communities, healthy homes, healthy people: initiating a research agenda on the built environment and public health. Am J Public Health 93(9):1446–1450

Tougwa FN (2015) The built environment discipline and its research at doctoral level. Research Paper, School of the Built Environment, University of Salford Manchester, United Kingdom

UN-Habitat (2013) State of the world's cities report 2012/2013: prosperity of cities, Earthscan, London

UN-Habitat (2016) World cities report 2016, urbanization and development: emerging futures, United Nations, Human Settlements Programme (UN-Habitat), Nairobi, Kenya

van den Berg L, Braun E, van der Meer J (eds) (2007) National policy responses to urban challenges in Europe. Ashgate, Aldershot, UK

Watson V (2016) Locating planning in the New Urban agenda of the sustainable development goal. Plan Theory 15(4):435–448
Wilson WH (1994) The city beautiful movement. Johns Hopkins University Press, Baltimore

The Built Environment as a Value Chain Process

Innocent Chirisa and Zebediah Muneta

Abstract The built environment and the construction industry are the key sectors in a functional economy as many economic activities take place in these structures. In functional economies, the main priorities are given to infrastructure development as it is one of the key drivers towards economic development and performance. There are many initiatives in the construction industry that creates less wasteful and operationally efficient building lifecycles through technological uptake, Internet of things and sustainable construction design. Importance of the built environment and the construction industry for triple bottom line of sustainability cannot be ignored since the built environment and construction have high economic importance and evident environmental and social impacts. Every business needs premises to operate and the built environment has significant direct and indirect impacts on social wellbeing and the livelihoods and affluence of local communities and individuals. Therefore, built environment has a significant role in sustainable development and significant potential also in shared value creation thus throughout the development process. This paper is a desktop research based on available literature paying particular attention to the need of investing in infrastructure in developing countries. The basic idea of this paper is to understand shared value creation in built environment context and the wellbeing of the society, users and residents.

Keywords Construction · People · Real estate · Value · Economy · Technology

I. Chirisa
Zimbabwe Ezekiel Guti University, Bindura, Zimbabwe

Department of Urban and Regional Planning, University of the Free State, Bloemfontein, South Africa

Z. Muneta (✉)
Department of Architecture and Real Estate, University of Zimbabwe, Mt. Pleasant, Box MP 167, Harare, Zimbabwe
e-mail: muneta.zeb@gmail.com

C. Chavunduka and I. Chirisa (eds.), *New Urban Agenda in Zimbabwe*,
Advances in 21st Century Human Settlements,
https://doi.org/10.1007/978-981-97-3199-2_2

1 Introductory Overview

The built environment and the construction industry are key largest sectors of today's global economy, representing 13% of GDP and employing 7% of the world's working-age population. Due to rapid urbanization in Africa, there is so much pressure on available infrastructure, warranting concern on the part of many African governments (Andersson and Andersson 2019). Urban population is growing rapidly, as a result, the size of the built environment is set to double, putting increasing pressure on urban systems such as water, energy and waste networks, and infrastructures need constant and continuous upgrade. Porter (1985) has defined a shared value process or chain as policies and operating practices that enhance the competitiveness of a company while simultaneously advancing the economic and social conditions in the communities in which it operates. Hence the importance of the built environment and the construction industry for triple bottom line of sustainability cannot be ignored since companies operate in these environments. The built environment and construction sector have high economic importance and evident environmental and social impacts. Every business needs premises to operate and the built environment has significant direct and indirect impacts on social wellbeing and the livelihoods and affluence of local communities and individuals (Gibberd 2020). Therefore, built environment has a significant role in sustainable development and significant potential also in shared value creation.

This global extension of technological advancement has contributed to value enhancement in the built environment since it has led to vertical chains and prioritization across different countries and has led to an increasing trade with inputs, intermediate goods and final products through sustainable infrastructure development. It is accompanied by growing transport and marketing activities, interregional and intraretinal (World Bank 2007). However, not all researches done in this context explore the contribution of value chain process in the built environment; hence, it became a field of interest, especially in the developing countries. Integrating the built environment in value chain in developing countries has also become a focus of this research, referred to as greening the value chain in property investment and development (UNEP 2019). Value chains are embedded in the built environment because economic activities and particularly agricultural production for example food or energy crops are based on environmental resources which is the land as it is a key factor of production. The sustainable use of land is based on the results of value chain analysis providing information on the input–output flows of products, but also on the effects of food, social and environmental standards implemented in different certification schemes. As the environmental value chain moved into the focus of public interest, the perception of consumers has increased and the environmental impact of products has become a major aspect of environmental policy programmes (Huws 2002).

Although recent decades have seen many improvements in the energy efficiency of buildings and the liveability of cities, the built environment we live in today continues to be designed around the linear 'take-make-dispose' model, in which materials are sourced, used and then disposed of as waste. The built environment, or the human-made environment, is the space that we live, work and recreate on a day-to-day basis (Moreno et al., 2020). From being born in hospitals, learning in schools, working in offices and living at home, almost all human activities are concentrated into different structures within the built environment (Couto et al., 2016). Globally, the construction industry generated $1.39 T in revenue in 2018 (Deloitte 2019) and a total built real estate valuation of $200 T (Savills 2019). Despite being the basis of our daily livelihoods, the construction, maintenance, use and demolition of properties are one of the most important sectors from the perspective of resource consumption and waste generation (Gibberd 2020). This means that virgin land value is increased through construction activities as it is recognized at the date of valuation by property valuers. In order to undertake such an analysis and consider these trade-offs, three questions must be applied to understanding natural resource use and environmental impacts along the construction value chain: (1) What is being built and where; (2) How much is being built; (3) How it is being built.

Desktop research was adopted in addressing the missing link between the built environment and the value chain process. Relevant information was identified from existing literature linking the built environment, technology and value chain management. Primary and secondary data were gathered from published articles from Google Scholar and annual real estate reports that are in line with smart construction activities in the world.

2 Theoretical Framework

Michael Porter's value chain, that is a strategic business planning tool used to identify where competitive advantage arises in a business, is the study's main theory. Every process from the beginning to delivery tracks the impact on a product or service. Five main activities and four support activities group the nine main stages of the value chain, and these include inbound logistics, operations, outbound logistics, marketing and sales, and service (UN-Habitat 2015). The value chain idea is based on the process views of the associations, the idea of a system consisting of subsystems, with inputs, transformation processes and outputs, that is to say the manufacturing (or services) organization. Inputs, transformation processes and outputs, as well as the consumption of resources such as money, labour, materials, equipment, buildings, land, administration and management, are all included in the acquisition. Costs and profits are determined by how value chain activities are carried out. In the process of converting inputs to outputs, most organizations engage in hundreds, if not thousands, of activities. These activities can be classified as either primary or support activities, and they are required of all businesses in some way (Hart et al. 2019).

The underlying concept of value chain was subject to different influences and objectives as it was only centred on the continuous process and value addition in products (Gibberd 2020). The origin of value chain process is described in two different concepts; that is the French concept and Wallerstein's concept of a commodity chain and from both of these concepts different aspects have developed (Dias 2003). The key concept of value chain process has a rich literature that is being implemented globally up to extent of the built environment in which many economic activities take place (Porter 1985).

The French concept value chain is more centred on agricultural products, and the concept was used to gain a more structured understanding of economic processes within production and distribution systems for agricultural commodities (Couto et al. 2016). In the 1970s, Wallerstein (1974) developed the concept of commodity chain, embedded in world systems theory that is an elaboration of the resource dependency theory which is well known at the global scale. The concept of a commodity chain is the base for the further developed global commodity value chain process that is similar to that of the built environment (Manewa et al. 2016). It seeks to explain the dynamics of the distribution of value chain activities in a capitalist world economy. The main driver is the international division of labour between different regions due to varying labour intensities of production and manufacturing activities within a chain in the built environment. In order to maintain competitive advantage in the built environment, companies in the construction and property sector must constantly examine the value they create due to the increasing competition on unsaturated prices, exceptional products and customer loyalty. A value chain can assist a company in identifying inefficient areas of its business and then implementing strategies to optimize its procedures for maximum efficiency and profitability brought about by the value chain process (Gibberd 2020). In addition to ensuring that production mechanics are smooth and efficient, businesses must ensure that customers feel confident and secure enough to remain loyal.

3 Literature Review

The built environment impacts the wellbeing of the people in a country through creating a healthy, safe and sustainable living and working environments. It requires the collective effort of a wide range of actors in the construction sector, from design and engineering to resource production, transportation, distribution and maintenance of the properties so as to keep the value generated by these key actors (Gibberd 2020). In recent decades, the construction sector has taken initiatives to become more circular and more sustainable through value enhancement with the fast technological uptake in the built environment (Elizabeth and Campling 2017). Due to globalization, it has presented various opportunities in the construction industry, built environment and value chain thereof. However, challenges such as climate change, geopolitics and scarcity of resources will adversely impact a smooth transition towards sustainable practices and a future-proof built environment. Besides that, some stages can take place simultaneously or with a different order such as the property market and the construction industry involve a mix of processes and actors in different parts of the world such as the production and supply of construction materials for construction activities to take place which results in the increase of new supply of new properties in the property market. Simplification and generalization on the stages of the value chain are necessary while acknowledging that these may vary between and within types of construction, countries and regions (Hart et al., 2019).

Construction as a sector globally has the highest material consumption footprint, consuming almost 50% of the total material footprint across the global economy (SCP-HAT 2020). As outlined, this involves the use of a range of natural resources and results in many environmental impacts. However, the construction sector also contributes in a variety of ways to socio-economic outcomes and to meeting the Sustainable Development Goals, in particular SDG 11 which calls for safe and affordable housing; sustainable and affordable transport systems; and inclusive, gender-responsive and sustainable urbanization. It is therefore essential to analyse the socio-economic outcomes of the construction sector alongside the associated natural resource use and environmental impacts, to be able to balance any trade-offs.

Discussions on sustainability in the construction sector often focus on the third question of how things are being built, with an emphasis on circularity of materials, energy efficiency and material substitutes and innovations, as well as on the safety of constructions and their resilience to natural disasters and to the impacts of global warming (Gibberd 2020). By changing how constructions are built through sustainable materials and practices, the associated natural resource use and environmental impacts of the sector can be reduced (Manewa et al. 2016). While this is a critical question, it is also necessary to ask what is being built and where? For example, is it residential, commercial, industrial or public infrastructure? Is the sector working on new builds or renovations of existing construction? Is the housing being built affordable or in cities and regions that have the greatest need for development? What land is being used for new construction? (WEF 2016). For example, if is it fertile land converted from agricultural purposes, or land already occupied by informal settlements. Depending on what type of construction is being built and where, this can have different levels of impact on natural resource use and environmental impacts, as well as contribute in varying degrees to meeting the SDGs, thereby influencing how trade-offs are balanced.

3.1 Key Observations Along the Construction Value Chain

A variety of construction projects involve different actors at the global scale, residential buildings hold the largest portion of construction. In relative terms, residential buildings constitute the major part of constructions in North America and Europe. Infrastructure dominates in Asia–Pacific, Latin America, Africa and the Middle East; while non-residential buildings make the strongest contribution in Eastern Europe (HIS 2013). Overall, governments are key players in the design, execution and management of institutional buildings and infrastructure, often with the co-participation of the private sector. Private companies play a prominent role in the conception, execution and management of residential and commercial buildings (Table 2.1).

Table 2.1 Main types of construction and indicative key players in their design, execution and management

	Residential buildings	Institutional and commercial buildings	Specialized industrial construction	Infrastructure and heavy construction
Examples of type of use	Apartments, houses, dormitories, sheds, etc	Schools, hospitals, offices, shopping centres, retail stores, stadiums, etc	Chemical industry plants, power plants, etc	Roads, tunnels, bridges, railways, sewage systems, pipelines, etc
Design	Architects, engineers	Architects, engineers	Specialized engineers	Government, architects, engineers
Execution	Construction companies, builders	Construction companies, builders	Industrials corporation or specialized companies	Government agencies and construction companies
Management	Individuals, real estate company	Retail: company/ private owner. School, hospitals: local/national government	Industrial corporation	Construction companies

Source UNEP (2021)

The construction sector and the built environment have a linear value chain with used ending as a property or infrastructure that has a fundamental value. Once a property is demolished at the end-of-life stage of the value chain, the vast majority of the materials used in the construction end up as waste, and this is consistent across most geographical regions (WEF 2016). In the European countries, the construction industry creates one of the biggest waste streams, with construction and demolition waste accounting for approximately 25–30% of all waste generated and these material losses rising year on year. This waste consists of numerous materials including concrete, bricks, tiles, wood, glass, metals and plastics, many of which could be recycled (Iyer-Raniga and Huovila 2020). The volume of construction waste is also increasing in Africa, where it is estimated to constitute 20% of all solid waste. In many African cities, this waste is not recycled or sent to landfill, but disposed of through uncontrolled dumping or open burning that contributes to pollution and health hazards (Iyer-Raniga and Huovila 2020; Gibberd 2020). Construction models are also linear in Latin America and the Caribbean where construction processes and policies do not support reusing or repurposing materials; however, some recycling does take place in the informal economy (Moreno 2020; Iyer-Raniga and Huovila 2020). In the Gulf Cooperation Council countries, construction also accounts for 35–40% of all solid waste in the region, with the majority ending up in landfill (Al-Alawi 2020).

The global perspective of value chains in the built environment was described by Nazareth (2019) as a strategic network, in this case value chains process does not simply exist in a particular space but are built for the better purpose of the users and the evironment. Hart et al. (2019) viewed value chain as a strategic network among a number of independent business organizations, where network members engage in extensive collaboration. World Bank (2003) reported that a built environment market system consists of a multi-player, multi-function arrangement comprising three main sets of functions (core, rules and supporting) undertaken by different players through which exchange takes place, develops, adapts and grows. The variation in explanations reflects the evaluation of the chain concepts from the different strands of debate for example agribusiness systems and supply chain management, world systems theory, participatory appraisal and the value chain concept (Nazareth 2019).

In the last years, many descriptive handbooks for different aspects on value chain analysis have been published. Walters and Rainbird (2004) summarized value chain concepts of relevance to development in the built environment as it helps in improving the infrastructure value as well as attraction of possible foreign investors in the sector. Also, Schmitz (2005) described the value chain process for policy-makers and practitioners. The relevance of value chain processes in the development of policies that strengthen economic relationships between actors in the built environment. World Bank (2007) provided the methodologies for value chain promotion in developing built environment projects. The target groups are public agencies and practitioners whose participatory value chain approach gains more and more importance of participatory value chain.

The building processes are very complex and hence need more interaction and collaboration between actors in the industry. There are various opportunities in the built environment to eliminate process and market barriers with organizational innovation leading to new business models focusing on collaboration between the different building actors and/or new construction value-chain-related ecosystems. Various types of loose or formal collaboration structures for achieving high-energy-performing buildings are appearing in all countries and the experiences connected to them are setting the scene for further development (Piroozfar 2013). Due to globalization and Internet of things coupled with the technological changes and energy-market levels, new market players are emerging in the construction value chain. Examples include the widely discussed launch of Tesla's Powerwall to stimulate demand-side flexibility and Google entering the home market with Nest thermostat and smoke alarm. Innovation will occur within companies, but also within ecosystems concentrating on a specific phase in a buildings' lifecycle, for example, collaboration models between architects, contractors and investors during the construction phase; or building automation for existing households set-up and designed by an installer for a specific family during the operation phase.

Globalization trends have significantly increased the scale and complexity of firms experiencing internal and external uncertainties (WEF 2016). Holzwarth et al. (2019) have described internal uncertainties during project development due to technological risks and external uncertainties, cover price uncertainties, exchange rate fluctuations, as well as demand variations of properties on the market. In a fast-moving economic environment, it is necessary to have tools to evaluate potential outcomes of changes and to capture complex surroundings in a simplified model. Value chain process in the built environment is a meaningful instrument to analyse and benchmark the sector with others in the economy. The operation of the current systems explains adaptation strategies to construction and economic challenges to reach a new efficient demand of properties and the need for upgraded infrastructure in the country. Dias (2003) noted that risk assessment is another tool to reduce the challenges during building maintenance for example environmental hazards of production activities and their resulting costs so as to have a strong impact on the distribution of profit and information flows.

Also, the aspect of value chain development features prominently in development programming aimed at stimulating economic growth and increasing competitiveness in the built environment and construction sector. The approach challenges governments and civil society to look beyond individual actors, such as small-scale enterprises or cooperatives, when considering how to achieve the development goals of the infrastructure. PwC (2014) has reported that by focusing on the value chain and the links between the actors spread along it, development interventions can better identify common problems among actors in the chain and solutions that generate win–win outcomes. Improved chain relations and overall chain performance are expected to yield tangible benefits in terms of economic performance and, in some cases, poverty reduction.

Information and materials at the different stages of the value chain process have an impact on resource use and environment in which the infrastructure is located. For example, IRP (2020) reported on the operation and use of building and related gas emissions while another report provided an overview of materials used in construction. Information on the use of materials along the global construction value chain is also limited. Information on materials is generally organized separately based on the specific type of material such as steel and cement, and its use across many different sectors. There is little analysis and knowledge to-date that combines an overarching view of the extraction and processing of the many different materials used along the construction value chain specifically (Munaro et al., 2020). This may also be related to the general focus on energy efficiency over material use in the reports addressing construction as it has implication on the final value of the property.

A really sustainable value chain demands a profound understanding and insight into resource production and heritage, and a comprehensive and sound assessment of good governance, social consequences, working conditions and environmental management (Munaro et al. 2020). Actors in the value chain can use digitalization to report more transparently, demonstrate accountability and better manage risks. Traceability is vital even if the value chain isn't completely transparent, one doesn't know where one's resources come from or how they're made. Digital systems allow traceability of materials and products through the various life-cycle phases. In recent decades, tracing of value chain of items (partly) has been through paperwork and tags such as the Radio Frequency Identification (RFID) (Holzwarth et al. 2019). The emerging blockchain technologies are designed to establish a digital tag that helps to track resources across value chain transactions and conversions.

In the world, government is a major investor in the built environment as it has a long-term interest. Governments can have substantial effect in the property sector volumes and types of business not only by regulating financial and property markets indirectly, but also directly by acting as suppliers for large infrastructure projects (Holzwarth et al. 2019). In 2017, the World Bank reported that 83 per cent of total US$0.5 trillion of global infrastructure investments came from the public sector, including investments by government and state-owned companies. At the global levels, as do individual countries through their overseas development financing, international organizations and multilateral development banks play a major role in the financing of infrastructures. In 2018, overall development funding from countries and multilateral development institutions spent on infrastructural projects in overseas was US$77.6 billion (OECD 2020).

Governments often contribute directly or indirectly to the funding of commercial infrastructure projects to encourage investment, 'governments are frequently required to provide funding, guarantees or other types of financial support for infrastructure projects' (World Bank 2007). The approach for sharing these resources includes needs and expectations assessment, financial support opportunities and all activities related to their sharing. This includes expenditure on investment, prospective financial sources and legal requirements as a way of increasing value to the land in the country. A full customer-investor offer is the next ingredient in the value chain. Any activity that could have a positive effect on environmental protection programmes is represented. Its addition is to offer and activate marketing that encompasses both the promotion of the region and the present and planned system of communication with investors (Munaro et al. 2020; Holzwarth et al. 2019). It enables the presentation of the region's worth and offer. The environmental field encompasses educational and information campaigns and media, exhibitions and contacts with similar regions, regional promotion, events and campaigns and incentives for the natural environment initiatives.

The other aspect is the fundamental activities and the inclusion of customer service in the value chain process it helps to increase benefits for corporations and institutions and investment in environmental protection. They comprise organizational, legal, as well as social and living support activities. The key objective is to provide more enticing services in territorial facilities (environmental policy, waste management water and wastewater management plans, development and modernization plans) management of staff in various organizations, development of concepts and procedures based on the existing research and technology potential in the built environment (Al-Alawl et al. 2020). The 'environmental value chain' has been moved into the public interest, consumer perception has been increased, and the environmental impact of products has become an important part of environmental policy programmes in developing countries.

4 Discussion

It is recognized that the development of the value chain process considers primary and secondary aspects described below. New relationships will be needed at each stage of a project thus from the development initiation to the purchase of undeveloped land (Brohman et al. 2020). The key relationship between the value-enhanced land and the virgin land showed the investment in the built environment. Valuation is done in order to capture the value created by the built environment cycle. There will be need to update the calculation of the value of built environment assets (Brohman et al. 2020). Key influencers in addition to stakeholder groups in the development process are considered in all phases of the development, such as quantity surveyors, architects, engineers and real estate agents, who leverage value creation. They will play a key role in new circular investment models. Inbound logistics entails working with suppliers and includes all activities related to receiving, storing and disseminating inputs.

All activities required to convert inputs into outputs are referred to as operations (products and services). Outbound logistics entails all activities required to transport goods from one location to another (Brohman et al. 2020). After a product or service is sold and delivered, all activities required to keep it working effectively for the buyer are referred to as service. Procurement is the process of acquiring inputs, or resources, for a company. All activities involved in recruiting, hiring, training, developing, compensating and dismissing or laying off personnel are included in human resource management (Al-Alawi et al. 2020). The equipment, hardware, software, procedures and technical knowledge used in the firm's transformation of inputs into outputs is referred to as technological development. Infrastructure consists of functions or departments that serve the company's needs and connect its various parts such as accounting, legal, finance, planning, public affairs, government relations, quality assurance and general management.

The constructed environment puts a lot of demand on natural resources due to urbanization, especially in Africa. Their involvement is crucial in the transition to a circular economy (CE). From a comprehensive approach, however, research must be intensified, including how new business models can enable materials to improve their residual values. This study gives an integral and complete review of the approaches to the study and actions of the circular economy in the built environment. The necessity to clarify CE and circular business models as well as to increase integration within the construction value chain among stakeholders is stressed (Brohman et al. 2020). Government support, such as subsidies, legislation and tax incentives, is essential for decision-makers' strategic performance so that circular concepts may be introduced and buildings and the constructed environment made sustainable. In addition to how the role of key influencer will evolve, we examine how enhanced environmental and social change can be reflected by valuation approaches. The biggest part of construction is contained in residential buildings on a global level. Relatively speaking, the primary part of buildings in North America and Europe is residential building. Asia-Pacific, Latin America, Africa and the Middle East are dominated by infrastructures; in Eastern Europe, non-residential construction is the biggest contribution (Hartenberger et al. 2013).

An important aspect in value chain is capitalisation of land rent and building value. Once land has building capital fixed to it, any measurement of respective rents will be better or worse indications. Property sales do not occur with a frequency that allows for especially enlightening time series and does not distinguish between land price and building price. On the other hand, they are better indicators of total capitalized value than are tax assessment values that in some cases remain at the same nominal level over a long period (Porter 1985). The advantages with tax assessment values are that they exist every year and distinguish between land value and building value. In order to capture the advantages of both data sources and improve their weaknesses, tax assessment records have been used to obtain annual data with land-building separation and bills of sale have been employed to correct their poor reflection of 'market value'. The timing of building depreciation, rent gap expansion, steep rise in capitalized land rent and finally redevelopment, depends on a myriad of circumstances and decisions that defy being structured into a mechanistic model.

There are several new intelligent business models, from sustainable procurement through to the tracking of items across the value chain in the construction sector. Within the Belgian construction business, the Belgian company Werflink promotes economic, social and environmental reforms and it is an online network for sharing equipment, materials, resources, freight and facilities that building sites and organizations can share during construction process (Iyer-Raniga and Huovila 2020). Companies can trade, sell and share untapped building equipment, materials, resources and space for freight and storage via the platform (WEF 2016). The aim is to use the available resources intelligently at the local level by the construction industry. Such a platform vastly reduces waste from unused or unneeded materials, reduces the reliance on new equipment and promotes local collaboration between different stakeholders.

Chain governance often refers to the vertical coordination by firms in one node of the chain with firms in other chain nodes. Coordination can assume various modalities that include strategic alliances and contractual partnerships. These determine how product flows are regulated in terms of prices, quality, quantity and delivery specifications, among other aspects (Humphrey and Schmitz 2005). Governance structures are considered to have important consequences for the access of chain actors in developing countries to markets and the range of activities that developing-country actors can undertake. A wider framing of the governance concept includes legislative aspects that shape business interactions, such as food safety and environmental standards (Kaplinksy and Morris 2002; Tallontire et al. 2011). The concept of upgrading refers to the potential of businesses and producers in developing countries to improve their performance and obtain greater benefits from value chain participation.

Construction process and demolition of structures can be planned for lifespan and flexibility to prevent the demolition of buildings in connection with changing market requirements. Buildings were created to be half-life expectancy in the last century alone. Longevity should be seen slightly different from before. Long-term construction does not necessarily indicate that for hundreds of years, a structure will remain robust; rather it hinders the development of modular building components that can be employed during the expected lifetime of various buildings (Nazareth 2019). Smarter building use is still a very traditional environment in developing countries. There is still great opportunity for development in regard of intelligent business models and use models, both for building structures and their use after construction (Debacker et al. 2017). A number of new start-ups have in recent years been able to offer new services thanks to the improvement of digital technologies. A surge of new business models, supported by digital technologies, is blooming in its created environment. Some of these approaches are driven by platforms that help to unleash the untapped value of other wasteful assets when used appropriately. The Resolve Framework sharing component promotes a different approach to material use that has been made far more feasible thanks to rapid adoption of 4IR technologies.

Due to the significance of the sector for both GDP and employment, regulation of construction can be a politically important issue for governments and economically important for private sector actors. Regulation of the construction sector has the potential to be used by governments for political aims, such as appealing to certain constituencies or stakeholders. Governments may also choose to deregulate the construction sector or weaken planning codes as a level to stimulate the sector in order to drive economic growth. Key private sector players working in the construction sector may undertake lobbying activities or make political donations that seek to influence the way governments regulate the sector in ways that are favourable to private interests (TIA 2019). This can sometimes relate to influencing the building code in order to promote the adoption and endorsement of particular building materials or technologies that could influence the natural resource use and environmental impacts of the sector. In some cases, the construction industry can also be subject to corrupt dealings between private companies and politicians, resulting in regulations, safeguards or assessments overlooked to make way for developments (World Bank 2007).

Urban planning and regulation of the construction sector by governments have strong potential to influence the natural resource use and environmental impacts of the sector; however, the existence of such regulation, as well as its quality and the degree to that it is effectively implemented, is influenced by a number of factors and interests across different countries. As a result, urban planning practices and regulation of the construction sector are not always consistent or effective, especially in developing countries.

5 Conclusion and Recommendation

In conclusion, only a few contextual transitional pathways and intervention sites for stakeholders across the built environment supply chain, including asset owners, are shown in the preceding examples. For example, though the use of natural resources and environmental impacts occur at the manufacturing of construction material stage, the construction stage and the operation stage of the value chain; the solutions can be found at many different stages. The interactions within the systems mean that the operations and behaviours of actors at different stages of the construction value chain can have a significant influence on the operations and behaviours of actors at other stages of the value chain. For example, the actions of investors and planners can shape the practices of construction companies, which in turn shape the characteristics of buildings and infrastructure available to users. It is therefore necessary to apply a systems lens to the analysis of a sector to move beyond a siloed and disconnected analysis, and towards understanding how different drivers of a given sector such as institutions, regulation, demographics and economic factors—shape the operations of along the value chain. Value chain balance models are a useful instrument in assessing complex links, including risk assessment and theorization of games, between actors and the environment. In microeconomic value chain impact assessment, econometrical value chain analysis is widely used. It includes treatment effect and gravity modelling to evaluate the impact on household or country incomes of food, social and environmental standards and transaction costs. Global analysis of the commodity chain aims at identifying and measuring the power balance between the economic players involved.

Construction also plays an integral role in meeting Sustainable Development Goals. Construction is critical to achieving the human right to adequate housing; building essential infrastructure necessary to provide mobility, energy, drinking water and sanitation; as well as to building the commercial and industrial infrastructure necessary to support economic development, all the while providing opportunities for employment and decent work. The importance of construction to meeting the SDGs means that there may be some trade-offs when it comes to natural resource use and environmental impacts of the construction sector, though it is also imperative that the sector transitions as much as possible towards resource efficiency, circularity and a smaller environmental footprint. However, the analysis has also shown that not all constructions contribute to sustainable development, and some activity in the construction sector can even result in harmful socio-economic consequences, making the natural resource use and environmental consequences difficult to justify.

Governments exert significant influence along construction value chain compared to other sectors, the role of governments and multilateral organizations in shaping activity along the construction value chain is significant and occurs as the regulators of financial markets, the banking system and tax systems, governments influence how much and what type of constructions are built, especially for housing, particularly at the financing stage and property market stage of the construction value chain. As investors in the construction sector through the public procurement of buildings and infrastructure, governments can directly influence what is being built, how much is being built and how constructions are being built through the procurement criteria they apply and the vendors they choose to engage.

As urban and territorial planners, and regulators of the construction sector, governments also indirectly determine what is being built, how much is being built and how constructions are being. How governments regulate the construction sector through tools such as building codes and zoning laws can influence the operations of actors along the construction value chain, especially at the planning and design stages;

the construction material stages; the construction stages; and, as a result, the use stage. Through the above three levers, governments have a major influence along the construction value chain on shaping what is being built, how much is being built and how constructions are being built, and it is these three levers that governments already use a combination of when stimulating the construction sector to boost economic activity or promote recovery during times of economic downturn or crisis. Governments therefore have a strong opportunity to reduce the natural resource use and environmental impacts of the construction sector by using these three key levers to drive resource efficiency in the sector and ensure construction activity is directed towards meeting the 2030 Agenda for Sustainable Development.

References

Al-Alawi B, Cavallari G, Cindric H, Gargallo M, Hegazy N, Saqib S, Shaka H, Tibi G (2020) State of play for circular built environment in gulf cooperation council (GCC) countries. Final report October 2020, ARUP and United Nations One Planet Network Sustainable Buildings and Construction Programme. https://www.oneplanetnetwork.org/sites/default/files/gcc_final_201101.pdf

Andersson DE, Andersson AE (2019) Sustainability and the built environment: the role of durability, sustainability, MDPI. Open Access J 11(18):1–19

Brohman MK, Parent M, Pearce MR, Wade M (2020) The business intelligence value chain: data-driven decision support in a data warehouse environment: an exploratory study. In: Proceedings of the 33rd annual hawaii international conference on system sciences. IEEE, pp 10-pp

Couto J, Tiago T, Gil A, Tiago F, Faria S (2016) It's hard to be green: reverse green value chain. Environ Res 149:302–313

Debacker W, Manshoven S, Peters M, Ribeiro A, De Weerdt Y (2017) Circular economy and design for change within the built environment: preparing the transition. In: International HISER conference on advances in recycling and management of construction and demolition waste, pp 114–117.

Deloitte (2019) Global powers of construction 2018. Accessed 21 Oct 2020

Dias WPS (2003) Useful life of buildings. Department of Civil Engineering University of Moratuwa

Elizabeth Havice and Liam Campling (2017) Where chain governance and environmental governance meet: interfirm strategies in the canned tuna global value chain. Econ Geogr 93(3):292–313

Hart J, Adams K, Giesekam J, Tingley DD, Pomponi F (2019) Barriers and drivers in a circular economy: the case of the built environment. Procedia Cirp 80:619–624

Hartenberger U, Lorenz D, Lützkendorf T (2013) A shared built environment professional identity through education and training. Build Res Inf 41(1):60–76

HIS (2013) Global construction outlook: executive outlook. IISD 2019. Guest article: why is it so difficult to fund infrastructure development in Africa? Accessed 21 Oct 2020

Holzwarth V, Rhy Search BS, Schneider SJ (2019) Approaching data driven services in the built environment. In: 2nd smart services summit, 18–30

Humphrey, Schmitz (2005) Useful life of buildings. Department of Civil Engineering University of Moratuwa

IRP (2020) Resource efficiency and climate change: material efficiency strategies for a low-carbon future. In: Hertwich E, Lifset R, Pauliuk S, Heeren N (eds) A report of the international resource panel. United Nations Environment Programme, Nairobi, Kenya

Manewa A, Siriwardena M, Ross A, Madanayake U (2016) Adaptable buildings for sustainable built environment. Built Environ Project Asset Manag

Moreno Z, Dias KW (2020) UK construction sector reports sharpest rise in almost five years. Accessed 21 Oct 2020 https://www.ft.com/content/328fe46f-d2c6-44e3-975b-7549e5340ecd
Munaro MR, Tavares SF, Bragança L (2020) Towards circular and more sustainable buildings: a systematic literature review on the circular economy in the built environment. J Clean Prod 260:121134
Nazareth AP (2019) How close is the built environment to achieving circularity? In: IOP conference series: earth and environmental science, vol 225, no 1. IOP Publishing, p 012070
Piroozfar P (2013) Advancements of mass customisation in service and manufacturing industries: lessons for the built environment. In: Mass customisation and personalisation in architecture and construction. Routledge, pp 55–65
Porter ME (1985) Competitive advantage, Chap 1. The Free Press, New York, pp 11–15
PwC (2014) Real estate 2020: building the future. Accessed 22 Oct 2020. Available online at: http real estatepwc.co.za
Savills (2019) Strategic management: a methodological approach, 4th edn. Addison-Wesley. Reading Mass
Schmitz (2005) Overview of Europe's building stock—renovate Europe. Accessed 14 Oct 2020
TIA (2019) Transparency International Australia Submission (Amended) to Senate Finance and Public Administration References Committee Inquiry Into: Compliance by former Ministers of State with the requirements of the Prime Minister's Statement of Ministerial Standards,' Transparency International Australia. Accessed 23 Oct 2020
UNEP (2019) Sand and sustainability: finding new solutions for environmental governance of global sand resources. GRID-Geneva, United Nations Environment Programme, Geneva, Switzerland
UNEP (2021) Analysis of the construction value chain: understanding the value chain and identifying hotspots. Catalysing science-based policy action on SCP task group IRP-one planet network, Nairobi, UNEP
UN-Habitat (2015) International guidelines on urban and territorial planning. UN Human Settlement Programme. Nairobi, Kenya
Walters D, Rainbird M (2004) The demand chain as an integral component of the value chain. J Consum Mark 2(1):118–129
WEF (2016) Shaping the future of construction: a breakthrough in mindset and technology. Switzerland, Geneva
World Bank (2003) Public money for private infrastructure: deciding when to offer guarantees, output based subsidies, and other fiscal support. The World Bank
World Bank (2007) Construction, corruption, and developing countries. World Bank Policy Research Working Paper 4271

The Biophysical Environment: Key Ingredient in Shaping the Built Environment

Joyline Chamunorwa, Jeofrey Matai, and Innocent Chirisa

Abstract The chapter explores the role of the biophysical environment, which is the natural environment, in shaping the built (the human-cultural) environment. It argues that the major source of problems affecting the biophysical environment is a result of the exploitative relationship between the two environments in favour of the built environment, yet the biophysical environment has the potential to shape the built environment. A mixed-methods research approach was adopted in the study. Documentary review was used to collect secondary data from relevant sources. Findings reveal that the biophysical environment is a key component in shaping the built environment because it presents multiple opportunities and resources that are critical to the functioning of cities. The biophysical environmental aspects considered are water, geomorphological assets, climate, soil and ecological resources. It is concluded that although cities present multiple challenges to the biophysical environment, there are options for improved and sustainable built environments presented by the biophysical environment. Recommended are planning, design and management of statutes and frameworks for the built environments that have provisions for the incorporation of the biophysical environment in the preparation of plans that guide the development of the built environment.

Keywords Environmentalism · Design · Construction · Policy · Sustainability · Planning

---

J. Chamunorwa · J. Matai (✉)
Department of Architecture and Real Estate, University of Zimbabwe, Mt. Pleasant, Box MP 167, Harare, Zimbabwe
e-mail: jeofreym@gmail.com

I. Chirisa
Zimbabwe Ezekiel Guti University, 1901 Barrassie Rd., Off Shamva Road, Bindura, Zimbabwe

Department of Urban and Regional Planning, University of the Free State, Bloemfontein, South Africa

C. Chavunduka and I. Chirisa (eds.), *New Urban Agenda in Zimbabwe*,
Advances in 21st Century Human Settlements,
https://doi.org/10.1007/978-981-97-3199-2_3

1 Introduction

The built environment is considered to have significant negative impacts on the biophysical environment. Extensive literature on the relationship between urbanisation, a process that produces the built environment, and the natural environment show that the built environment impacts negatively on the biophysical environment by, among other things, destroying land cover, loss of biodiversity and compromising water quality (Dodman et al. 2013; Kharlamova et al. 2016; Opoku 2019). The built environment depends highly on the natural environment. For example, Bolund and Hunhammar (1999) argue that whilst humanity is increasingly urbanising, urban people still rely on rural environs for inputs and processing of the outputs from the cities. Similarly, the natural environment, through the integration of nature in the built environments has positive benefits such as the cooling effect of vegetation (Dodman et al. 2013; Wang et al. 2018). These, among other benefits derived from the biophysical environments by cities and their inhabitants, show how important the biophysical environment is to cities and city dwellers. Whilst the biophysical environment is key to the sustenance of cities, it also poses challenges to the built environment. The Secretariat of the Convention on Biological Diversity (2012) puts it that cities are experiencing challenges that include inter alia, shortage of natural resources, rising temperatures, frequent floods and heatwaves. These are aspects of the biophysical environment that pose devastating impacts on the built environment although it can be argued that the continued growth of the built environment is the reason behind these challenges (Kahn 1996; Opoku 2019; Sakalasooriya 2013).

Studies on the relationship between the built environment and the biophysical environment are biased towards the skewed relationship between the two, in favour of the built environments. For example, Haase (2009) focused on how urbanisation affects the water balance. Trotter et al. (1990) indicated that the rapid growth of the built environments impacts on the temperature whilst scholars such as Bai et al. (2017), Clement (2010) and Rees (1999) also show that the built environment and the activities associated with it pose challenges to the biophysical environment. In Zimbabwe, scholarly work also indicates that the built environment impacts negatively on the environment through, inter alia, increased exploitation and consumption of the biophysical assets (Chirisa 2013), reducing water quality and forests (Kibena et al. 2014; Nhapi et al. 2004) and increasing surface temperatures (Unganai 1997). In a way, this shows a consensus among scholars that the built environment presents challenges to the natural environment. Policy recommendations coming from these studies are inclined to the opportunities that the built environment presents for the protection, conservation and betterment of the biophysical environment (Bai et al. 2017; Kahn 1996; Opoku 2019; Pieterse et al. 2017; Rees 1999). This study acknowledges that the built environment presents opportunities for the protection and improvement of the biophysical environment, but also argues that the biophysical environment can be a key ingredient in shaping the built environment, particularly in the Zimbabwean context. An attempt to show that the biophysical environment can shape the built environment was made by Matai and

Zimunya (2020) but narrowly focused on the role of vegetation in modifying temperatures in urban local environments. This study, therefore, explores and explains how the natural environment can be used as an ingredient in shaping cities using Harare city as a case study. The selection of Harare city is on the basis of the multiple water bodies, wetlands, animal and vast plant species which the city is developed on, making it a study area with multiple and complex biophysical assets that can comprehensively address the research question.

The chapter is organised into six sections. Following the introduction is the methodology used to carry out the research. This is followed by the literature review and theoretical framework which positions the study among the exiting body of literature and the relevant theories. This will be followed by the discussion of findings and the conclusion and recommendations section.

2 Methodology

The purpose of the research is to examine the role that the biophysical environment can play in shaping the built environment. To realise this aim, an exploratory approach and a qualitative research method was adopted. Interviews were conducted with key informants who constituted purposefully selected experts in urban planning and design and environmentalists. Observations were also used to collect data that was useful to the study. This includes data on the physical environmental aspects of the city, environmental issues in the city that are instrumental in shaping the city environs. Secondary data in the form of the Master Plan, layout plans and layout reports, maps such as the Harare wetlands and topographic maps and newspaper articles were also used as data sources for the study through document analysis. Data was analysed through thematic narratives. Consent was sought before data collection and high levels of confidentiality of information from participants were ensured.

3 Literature Review and Theoretical Framework

This section reviews literature on the relationship between the biophysical environment and the built environment. Firstly, the section unpacks the key components of the biophysical environment and the built environment and provides a conceptual framework for the study.

3.1 The Biophysical Environment

The biophysical environment consists of the biotic, such as plants and animals, and abiotic, for example, rocks, soils and water that surround an organism or population, including factors that influence their existence, growth and evolution. The biophysical environment is made up of four parts that are the atmosphere, hydrosphere, lithosphere and the biosphere and ranges from the microscopic to the global extent (Natural Water Retention Measures 2016). The atmosphere contains gases such as nitrogen, oxygen and other elements that include water vapour, dust and ash particles. According to Doran and Zimmerman (2009), the atmosphere is a shielding layer that sanctions survival on the earth's surface. The composition of the atmosphere is affected by anthropogenic activities. For example, burning of fossil fuels increases the level of Green House Gases (GHGs) in the atmosphere resulting in the rise in temperature. Changes in land cover also contribute to rising temperatures in the earth's atmosphere (ibid).

The hydrosphere consists of all waters on the surface of the earth such as lakes, seas and also water under the earth's surface and above the earth's surface, for example clouds. Water can be found in various states: liquid, vapour or ice. The processes of evapotranspiration, condensation, infiltration, precipitation, run off and subsurface flow are regarded as the wheels in the hydrologic cycle that move water in and out of the hydrosphere (Della Sala 2018). Like the atmosphere, the hydrosphere is also affected by human activities. Kundzewicz (2008) argues that the increase in human population increases demand for water resulting in over extraction. It has been indicated that although per capita water use in some cities is decreasing, total water use is increasing due to increased use of water in outdoor activities. Other human activities that affect the hydrosphere include wasteful consumption of water, pollution, agro-industrial as well as processes of fossil fuel extraction (Della Sala 2018). Most of these activities are linked to the built environment.

The proportion of the atmosphere that sustains life is referred to as the biosphere (Wang et al. 2017). It is one of, among other spheres of the earth, where living things thrive and live. Others define the biosphere as the sum of all ecosystems on the earth (Vladimir 1998; Wang et al. 2017). The biosphere is said to consist of the lithosphere, atmosphere and the hydrosphere, entailing that sunlight and water are also part of the biosphere. The biosphere has been changed extensively by human activities through, for example, the exchange of species between ecosystems resulting from increased globalisation (Ludwig and Steffen 2018). Similarly, humans are altering the biosphere through, among other activities, urbanisation and agricultural practices. Considering that there is an intricate relationship between the biosphere and humans (Verma 2018), changes in the biophysical environment will affect the survival of human beings.

The lithosphere, as another component of the biophysical environment, refers to the outer rigid shell of the earth, hence, it is sometimes called the earth's skin. Thus, soils and rocks constitute the lithosphere. It corresponds with the topographic surface and is highly heterogeneous. Like other components of the biosphere, the lithosphere has been changed and Oguchi (2020) states that the pace of change is increasing. Some of these changes are linked to the built environment (Artemieva 2011). The built environment, for example, takes up space and modifies the earth's surface, hence, impacts on the lithosphere.

3.2 The Built Environment

The built environment constitutes all aspects of our lives, incorporating the buildings, the systems that distribute water and electricity, and the roads, bridges, as well as the system of transportation that enables city dwellers to move from place to place. Because it is made up of modified structures to provide people spaces for living, working and recreation; it is sometimes called the man-made environment (Kaklauskas and Gudauskas 2016). The creation of the aforementioned sub-systems of the built environment and spaces requires enormous quantities of materials from the biophysical environment (McDonald et al. 2013; Opoku 2019). The definition of the built environment has been expanded to include access to health, food, community gardens, walkability and likability (Carmona 2019; Montgomery 1998). This implies that the built environment goes beyond the physical aspects to include services and activities therein. This also means that the built environment is created by day-to-day human activities.

3.3 The Biophysical Environment and the Built Environment: The Nexus

The similarity between the biophysical environment and the built environment is that parts or components of these environments interact. For example, in the built environment, an improved transportation component determines the nature and extent of growth of a city by enabling city dwellers to stay further away from places of employment (Bai et al. 2017). Components or parts of the biophysical environment also interact. For instance, the atmosphere creates a climate that accommodates a diverse range of plants and animals. Changes in one component of the environments will impact on the other components. Besides the interactions within the environments themselves, there are interactions between the biophysical environment and the built environment. The changes in either of the two environments have effects and implications on the other environment. The sprawling nature of cities increases demand and construction of roads and parking spaces. Run off from the parking spaces and roads carries pollutants that contaminate the rivers, lakes, streams and oceans (Kahn 1996). The built environment creates hard impervious surfaces that replace forests such that rain can no longer seep into the ground to recharge aquifers. The occurrence of natural disasters emanating from the changes in the biophysical environments also impact on the built environment mostly by destroying the building structures and infrastructure, for example in instances of floods and tsunamis.

The interactions between the two environments can influence the assets and components of the other environment. However, it is generally acknowledged that it is the built environment and the activities within it that have more negative effects on the biophysical environment. Seto et al. (2013) and the Secretariat of the Convention on Biological Diversity (2012) identified trends that link urbanisation to the biophysical environment. These are based on how the transformations taking place in the cities result in changes in the biophysical environment and vice versa. One of the trends, which is of critical importance to this study, is that of the unproportional increase in urban population and the physical expansion of cities. The total area to be covered by urban areas is expected to triple by 2030, whilst urban population is expected to double during the same time (Secretariat of the Convention on Biological Diversity 2012). This indicates that the built environments are expanding faster than the population inhabiting them. Contrary to the view that compact cities reduce demand for raw materials and infrastructure (Seto et al. 2013), the expanding cities increase demand for water and land and according to the Secretariat of the Convention on Biological Diversity (2012), this has knock on effects on the components of the biophysical environment.

Overall consumption patterns by urban inhabitants are also rising consequently placing greater strain on resources. The lifestyles of the people, especially those who inhabited the areas before they urbanised, change. Besides turning to the natural environment for survival, for example through wetland cultivation to supplement food supplies, cutting down of trees for firewood and sand mining among other biophysical environment components, patterns associated with urbanisation, especially the increasing demand for residential energy and water place pressure on ecological services (Hubacek et al. in Seto et al. 2013). Besides increasing demand for natural resources, cities are also drivers of climate change, for example, dependence on motorised vehicles for movement increases the emission of greehouse gasses which results in the depletion of the ozone layer and the consequential effects on climate. Urbanisation is associated with considerable land cover changes that impact on temperature and precipitation within and around urban areas. Heymans et al. (2019) state that 80% of greenhouse gas emissions that cause climate change are accounted for by the built environment, and that the reduction of vegetation and increased impermeable surfaces contribute to heat waves that have adverse effects on public health. The physical alterations to land use patterns, degradation of habitats and fragmentation profoundly change the natural environment posing threats to species diversity and the ecosystem (Opoku 2019). All these changes in the built environment have profound effects on the biophysical environment. This understanding of the effects of the built environment on the biophysical environment makes it clear that policy making and practice in the creation and management of the built environment should integrate the biophysical environment. To this end, significant efforts have been made towards integrating the environment in the planning, design and management of cities which are the built environment (Heymans et al. 2019; Opoku 2019; Sakalasooriya 2013; Secretariat of the Convention on Biological Diversity 2012; Seto et al. 2013). As much as this is a critical step towards the protection and conservation of the biophysical environment, the question of how the biophysical environment can be used as an ingredient to shape the built environment remains scantly answered, in most instances, indirectly.

Based on the discussion provided above on the relationship between the built environment and the biophysical environment, it is clear that there are dependencies between the two environments. Considering the consumption by the inhabitants of the built environment (Seto et al. 2013), it can be argued that instead of depending on the natural environment outside the city for supplies of food, production can be done within the boundaries of cities. Müller et al. (2013) indicated that urban green spaces have historically been significantly providing agricultural produce, fish and game among other crucial food items. In contemporary cities, urban residents produce approximately 15–20% of the world's food (Müller et al. 2013). Further to this, 90% of vegetables consumed in Dar es Salaam originate from the urban and peri-urban areas, while 58% of rice consumed in Hanoi, whose population is approximately above 4 million, is produced within the jurisdiction of the city (Müller et al. 2013). Such figures show that it is possible to shape the built environment by considering components of the biophysical environment during the planning and design stage of the built environment. This will not only lower the costs of managing the environment and of food provision, but also minimise the ecological footprint of cities.

Kaklauskas and Gudauskas (2016) assert that the built environment is developed to meet the requirements of the residents. These needs can either be physical or social. To achieve this, the built environment should be aesthetically attractive, be accessible with functional and good quality infrastructure with comparatively cheap and comfortable dwellings. The biophysical environment can play a big role in ensuring that these needs and expectations of city dwellers are satisfactorily met if planners and designers make use of the key components of the biophysical environment in shaping the built environment. For example, the use of local material in the construction of buildings can make housing affordable and cheaper to maintain. Similarly, integrating nature in the form of parks, trees along major streets (Carmona 2019; Cheshmehzangi and Butters 2017; Ness and Xing 2017) can make them aesthetically attractive and cooler and also reduce demand for energy for cooling and heating (Matai and Zimunya 2020; Wang et al. 2018).

The discussion above, on the biophysical environment, the built environment and the relationship between the two environments shows that there are opportunities for planning, designing and managing the built environment using the biophysical environmental components as the guiding framework. For example, one can focus on the following questions in a bid to do so: how can cities be designed to ensure that the hydrosphere is not affected and to capitalise on the benefits provided by the hydrosphere to improve the quality of life of urban dwellers? What aspects of the biosphere can be incorporated in the planning, design and management of the built environment to ensure that the built environment has less energy demands, and is aesthetically attractive and sustainable? How can the lithosphere be used to guide and shape the form of cities and to ensure that the day to day needs of urban inhabitants such as food are adequately provided for and how can the atmosphere be maintained in a clean state with suitable proportions of water vapour, oxygen and other gases to guarantee comfort and minimise externalities from the poor material composition of the atmosphere? Whilst these questions can be raised independently, using the components of the biophysical environments to shape the built environment requires that the components are considered holistically in a systems approach.

4 Theoretical Framework

The study is grounded in the theory of sustainability. The sustainability theory emphasises the importance of using natural resources to promote human welfare while respecting the carrying capacity and reproductive capacity of the natural environment (Caldwell 1998; Geerlings and Vellinga 2017; Khalili 2011). The ultimate goal of sustainability is the achievement of a balance between development and the environment which according to Keeley and Benton-Short (2018), is only achievable through the integration and appreciation of three inter-linked dimensions; social, economic and environmental issues in decision making. Striking a balance between the social, economic and environmental issues for sustainability should also consider time, that is, the present and future generations (World Commission on Environment and Development 1987). Sustainability is also attained when the resources available are equitably shared (Khalili 2011). This generally implies that sustainability is both intra and intergenerational.

In the context of this study, the theory of sustainability provides a basis for interventions in the built environment to ensure that natural resources from the biophysical environment are used sustainability to ensure use and enjoyment within and between generations. This implies that, in shaping the built environment, serious considerations should be put on how the development and growth of cities can contribute to the improvement and not harming the biophysical environment. Whilst this is the key argument of the sustainability theory, it should be viewed, from the perspective of this study that the biophysical environment also plays a key role in ensuring the sustainability of the built environment. Therefore, the theory in this study is critical in informing the possible measures that can be adopted to shape cities that are sustainable by capitalising on the biophysical environment.

5 Discussion of Findings

This section presents the findings of the study. The findings are presented along thematic areas on the basis of the characteristics of components of the biophysical environment identified in Harare and how these can be used to inform the planning, design and management of the city. An evaluation of the current practices with respect to the relationship between the biophysical environment and the built environment is also given. Put differently, the presentation of findings also provides an analysis of the existing situation in Harare regarding how the city's planning and design is informed by the biophysical environment.

The Biosphere

The City of Harare is resourced with a diversity of elements of the biosphere that include plants, animals, wetlands, forests and aquatic life. In 2010, Harare had a tree cover of 942 hectares, representing approximately 1.1% of the city's total area. These are distributed across the city in the form of botanical gardens, wildlife parks areas, forests and public parks (Global Forest Watch n.d). Animals and birds are also found in the woodlands, parks and forests among other places within the city. However, by 2020, Harare lost 4.77 hectares of tree cover, equivalent to 1.48 kt of carbon dioxide of emissions (ibid). This significant change is attributed to the growth of housing estates that have seen the town sprawling, informal housing developments, illegal cultivation and wanton cutting down of trees as people, particularly the poor, resort to firewood as a source of energy. In this regard, it can be deduced that the planning and design of Harare is not capitalising on this element of the biophysical environment, which is a key ingredient in addressing climate change. The tree cover, as indicated earlier in the literature review, plays a critical role in sequestrating carbon from the atmosphere, thereby improving the quality of the built environment by ensuring that the air in the city is clean. Findings from experts in this regard reveal that whilst the planning systems, which constitute planning and designing, are in place to guide and regulate development, there is a general lack of political will and resources to enforce the law as specified in the relevant Acts such as the Regional Town and Country Act, Urban Councils Act and the Environmental Management Act.

Harare city is also characterised by several pockets of classified wetlands that are dotted across the city as shown in Fig. 1. These wetlands provide for a diverse array of species and ecosystem services for the inhabitants of Harare. According to BirdLife Data Zone (2015), wetlands in Harare have over 240 bird species recorded including some rare species. However, there is empirical evidence that the wetlands are being encroached by human activities, chief among them being housing development as indicated in the study by Sithole and Goredema (2013).

Linked to the wetlands are rivers and dams. These form part of the hydrosphere of Harare and are distributed across the city. Most of these flow outside the city centre as shown in Fig. 1. The major source of water for the city is Lake Chivero which Nhapi (2009) states that water extracted from the dam for use returns to it as sewerage effluent. In that regard, the built environment of Harare negatively affects the hydrosphere by polluting the water and also by extracting approximately 30% of the water. Considering the water problems that are affecting the city, both in terms of quantity and quality, it can be seen that the current planning and design practices in the city do not capitalise on the hydrosphere element of the city's biophysical environment. Evidence from analysis of the river system and the city shows that the city sits on the watershed, a location that exposes the rivers that drain into the lake Chivero and other dams that supply water to the city to pollution from the waste generated in the city. This is an indication of how the planning and design considerations of the city have failed to capitalise of the rich hydrosphere of the city. Thus, with abundant surface water sources that could significantly meet the water

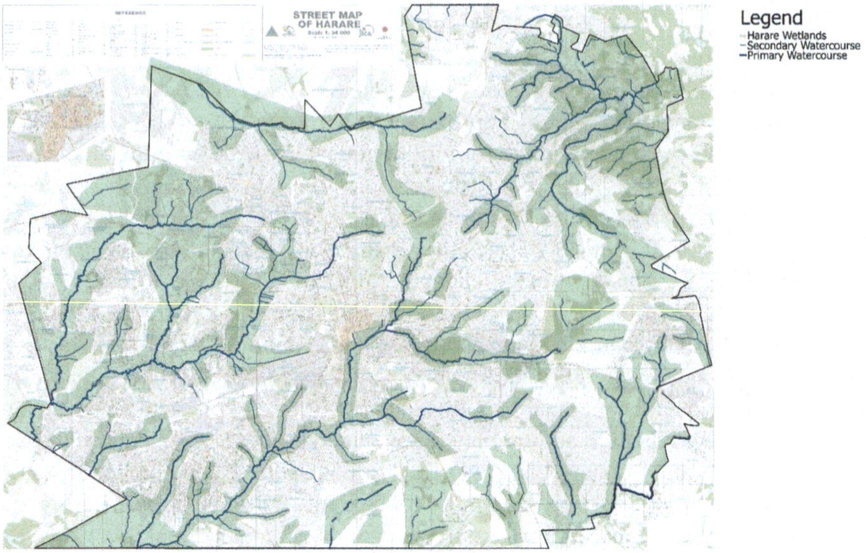

Fig. 1 Harare wetlands and rivers. *Source* UNDP 2020

demand for the city, careful consideration of location of the city that recognises the water basin could have made the city's water challenges less than they are today.

Harare is also endowed with underground water which is accessible through the drilling of boreholes at an average depth of fourty metres. Most new settlements in the city do not have access to piped water, whilst earlier developments experience erratic water supply. In response to this situation, some residents, organisations and institutions across the city are sinking boreholes. Although there is no statistical data on the number of boreholes across the city currently, observation shows that a significant number of households in the city have sunk and are sinking boreholes on individual plots to meet their needs for clean, safe and adequate water supplies. As a result, there is over extraction of underground water. This has negative impacts on the quantities and level of underground water. Agricultural practices are also common in Harare some of which make use of fertilisers and pesticides that in turn, pollute the underground water sources. Findings from interviews with key informants and environmentalists indicate that the continued reliance on underground water in the city, although not quantified, can result in scarcity of water in the city. The scenario presented here indicates that the planning of Harare is not capitalising on the hydrosphere, specifically abundance of underground water. A systems approach that considers urban development, population growth and the underground and surface water sources can be used to inform the planning and design of the city. Planned decentralisation of water supply can be used to resolve water challenges considering the abundance of both surface and underground water sources across the city.

The earth's surface, the lithosphere, of Harare is undulating with the greater part of the land sloping gently and developable. The general topography of the city presents opportunities for easy drainage of rain water, water and sewer reticulation. The orientation of road infrastructure is also made easy because of the terrain although there are parts of the city where this is not easy. The lithosphere characteristics also present opportunities for the construction of multi-storey buildings in most parts of the city. The soils that make up the upper crust of the lithosphere in Harare are also rich in texture and nutrients and therefore have potential for high agricultural productivity. As such, the lithosphere also presents opportunities for ensuring food security of Harare residents. One expert indicated that Harare has potential to feed a large proportion of the city's population if urban agriculture is well integrated in the planning and designing of the city. Reasons given were linked to the suitability of the soils and the availability of water from underground, wetlands and multiple streams and rivers that can be harnessed for irrigation at lower costs as well as reliable rainwater that can also be harvested. In addition to the opportunities presented by the lithosphere for development, there are several sites within the city where gravel and sand for construction of roads and houses can be mined. These are key ingredients from the biophysical environment that play a critical role in shaping the built environment, its functionality and quality.

The climate of Harare, which is characterised by a moderately wet climate, and the presence of large plot sizes in residential areas, and several large open spaces within the boundaries of the city provide conditions that are good for urban food production. In addition to the climatic conditions, Harare is also characterised by vleis that are seasonally waterlogged on the clay and sandy soils. The vleis play a critical role in retaining moisture for a long period through to the next wet season. Such vleis provide an opportunity for communities to plant crops early and late, enabling farmers to harvest twice a year using the same piece of land. Thus, vleis provide suitable conditions for gardening. As far as urban agriculture is concerned, the relatively large plots allow communities to practice farming on the plots around their houses (Mbiba 1999). This goes a long way in meeting household demand for leafy vegetables. Whilst the vleis provide suitable conditions for urban agriculture, they affect built structures, sewer and sewer pipes because of their soil characteristic of expanding when wet and contracting when dry.

Based on the key components of the biosphere that characterise Harare, there are opportunities presented by the biophysical environment to guide the planning, design and management of cities to make them more sustainable either through retrofitting, redevelopment programmes or in the creation of new places within the city. One of the key challenges that Harare city is experiencing is shortage of water. This shortage is a result of issues that include increased population against an unproportional rate of water infrastructure development. Making use of the hydrosphere, and the hydrological system in general, the city receives an average annual rainfall of 840 mm (World Climate Guide n.d). Whilst this can change due to climate change, creation of systems and infrastructure for harvesting water can go a long way in addressing the water challenges in the city. This entails the planning, design and management of the built environment, Harare in this case, which should make provision for the requisite infrastructure, including decentralised and strategically positioned storage tanks and dams to harvest the water.

The biophysical environment for Harare provides a suitable environment for the survival of plants. This provides opportunities for greening the city and for food production. However, the absence of a framework, legal or policy, to guide greening and urban agriculture in Harare creates a gap as there is a likelihood of reluctance by the authorities to implement them. In a way, the biophysical environment has potential in this case, to create a city that is less dependent on energy for cooling and heating through the integration of trees in the city environment (Matai and Zimunya 2020; Wang et al. 2018). It also presents an opportunity for the creation of a city that can produce food for the inhabitants, thereby reducing the ecological footprint of the city. This also means that zoning considerations during Master Planning should consider spaces for tree planting and for agricultural activities. This has an impact on the form and structure of cities as well as approaches to the management of cities as new codes should be developed to manage such land uses.

6 Conclusion and Recommendations

Findings reveal that the biophysical environment for Harare and immediate environments provide opportunities for shaping the built environment. Thus, the biosphere, lithosphere, hydrosphere and atmosphere for Harare provide a conducive environment for the creation of a city that is sustainable, inclusive and resilient. It can therefore be concluded that with careful planning that considers the suitability of the built environment in a holistic manner, the biophysical environment can influence the manner in which the built environment can be planned and designed as well as managed. The recommendations made are that the government should craft policies that encourage consideration of the biophysical conditions of urban areas in planning, shaping and managing the built environment. This means that planning frameworks should be revised to include issues that go beyond the traditional site analysis.

References

Artemieva I (2011) What is the lithosphere? Lithosphere. 1–14. https://doi.org/10.1017/CBO9780511975417.003

Bai X, McPhearson T, Cleugh H, Nagendra H, Tong X, Zhu T, Zhu YG (2017) Linking urbanization and the environment: conceptual and empirical advances. Annu Rev Environ Resour 42:215–240. https://doi.org/10.1146/annurev-environ-102016-061128

BirdLife Data Zone (2015) Wetland conservation and environmental education in the heart of Harare. http://datazone.birdlife.org/sowb/casestudy/wetland-conservation-and-environmental-education-in-the-heart-of-harare

Bolund P, Hunhammar S (1999) Ecosystem services in urban areas. Reg Econ 29(2):293–301. https://doi.org/10.1017/S174217051300046X

Caldwell LK (1998) The concept of sustainability: a critical approach, Springer, Dordrecht, pp 1–15. https://doi.org/10.1007/978-94-017-1337-5_1

Carmona M (2019) Place value: place quality and its impact on health, social, economic and environmental outcomes. J Urban Des 24(1):1–48. https://doi.org/10.1080/13574809.2018.1472523

Cheshmehzangi A, Butters C (2017) Designing cooler cities: energy, cooling and urban form—The Asian perspective. In: Palgrave series in Asia and pacific studies. Springer

Chirisa I (2013) Politics in environmental stewardship in Zimbabwe: reflections on Ruwa and Epworth. Afr J Hist Cult 5(1):13–26. https://doi.org/10.5897/AJHC11.026

Clement MT (2010) Urbanization and the natural environment: an environmental sociological review and synthesis. Organ Environ 23(3):291–314. https://doi.org/10.1177/1086026610382621

Della Sala DA (2018) Freshwater and global change: wellspring of life. Encyclop Anthropo 1–5:21–24. https://doi.org/10.1016/B978-0-12-809665-9.05876-6

Dodman D, McGranaham G, Dalal-Clayton B (2013) Integrating the environment in urban planning and management. https://stg-wedocs.unep.org/handle/20.500.11822/8540

Doran PT, Zimmerman MK (2009) Examining the scientific consensus on climate change. Eos 90(3):22–23. https://doi.org/10.1029/2009EO030002

Geerlings H, Vellinga T (2017) Sustainability. In: Ports and networks: strategies, operations and perspectives. Taylor and Francis, pp 296–314. https://doi.org/10.4324/9781315601540

Global Forest Watch (n.d.) Harare, Zimbabwe deforestation rates and statistics. https://www.globalforestwatch.org/dashboards/country/ZWE/2/1/

Haase D (2009) Effects of urbanisation on the water balance—A long-term trajectory. Environ Impact Assess Rev 29(4):211–219. https://doi.org/10.1016/j.eiar.2009.01.002

Heymans A, Breadsell J, Morrison GM, Byrne JJ, Eon C (2019) Ecological urban planning and design: a systematic literature review. Sustainability (Switzerland) 11(13):1–20. https://doi.org/10.3390/su11133723

Kahn ME (1996) Green cities: urban growth and the environment. Brookings Institution Press. https://doi.org/10.1007/978-981-15-4386-9_12

Kaklauskas A, Gudauskas R (2016) Intelligent decision-support systems and the Internet of Things for the smart built environment. In: Start-up creation: the smart eco-efficient built environment. Elsevier Ltd. https://doi.org/10.1016/B978-0-08-100546-0.00017-0

Keeley M, Benton-Short L (2018) Urban sustainability in the US: cities take action. In: Urban sustainability in the US: cities take action, pp 1–337. https://doi.org/10.1007/978-3-319-93296-5

Khalili NR (2011) Theory and concept of sustainability and sustainable development. In: Practical sustainability. Palgrave Macmillan, New York, pp 1–22

Kharlamova MD, Mada SY, Grachev VA (2016) Landfills: problems, solutions and decision-making of waste disposal in Harare (Zimbabwe). Biosci Biotechnol Res Asia 13(1):307–318. https://doi.org/10.13005/bbra/2034

Kibena J, Nhapi I, Gumindoga W (2014) Assessing the relationship between water quality parameters and changes in landuse patterns in the Upper Manyame River, Zimbabwe. Phys Chem Earth, Parts A/B/C 67–69:153–163. https://doi.org/10.1016/J.PCE.2013.09.017

Kundzewicz ZW (2008) Hydrosphere. Encyclopedia of ecology, five-volume set, pp 1923–1930. https://doi.org/10.1016/B978-008045405-4.00740-0

Ludwig C, Steffen W (2018) The 1950s as the beginning of the Anthropocene. Encycl Anthropocene 1–5:45–56. https://doi.org/10.1016/B978-0-12-809665-9.09940-7

Matai J, Zimunya W (2020) Connecting biophilic design to neighbourhood energy use optimisation in rural towns: The case of Birchenough bridge. J Urban Syst Innov Resilience Zimbabwe 2(1):40–60

Mbiba B (1999) Urban agriculture in Harare: between suspicion and repression. University of Sheffield

McDonald RI, Marcotullio PJ, Guneralp B (2013) Urbanisation and global trends in biodiversity and ecosystem services. In: Elmqvist T, Goodness J, Marcotullio PJ, Parnell S, Sendstad M, Wilkinson C, Fragkias M, Guneralp B, McDonald RI, Schewenius M, Seto KC (eds) Urbanization, biodiversity and ecosystem services: challenges and opportunities: a global assessment. Springer, pp 31–51. https://doi.org/10.1007/978-94-007-7088-1_23

Montgomery J (1998) Making a city: urbanity, vitality and urban design. J Urban Des 3(1):93–116

Müller N, Ignatieva M, Nilon CH, Werner P, Zipperer WC (2013) Patterns and trends in urban biodiversity and landscape design. In: Elmqvist T, Goodness J, Marcotullio PJ, Parnell S, Sendstad M, Wilkinson C, Fragkias M, Guneralp B, McDonald RI, Schewenius M, Seto KC (eds) Urbanization, biodiversity and ecosystem services: challenges and opportunities. Springer, pp 123–176

Natural Water Retention Measures (2016) Biophysical environment I natural water retention measures. http://nwrm.eu/concept/3944

Ness DA, Xing K (2017) Toward a resource-efficient built environment: a literature review and conceptual model. J Ind Ecol 21(3):572–592

Nhapi I (2009) The water situation in Harare, Zimbabwe: a policy and management problem. Water Policy 11:221–231. https://www.nationalgeographic.org/encyclopedia/hydrosphere/print/

Nhapi I, Siebel MA, Gijzen HJ (2004) The impact of urbanisation on the water quality of Lake Chivero, Zimbabwe. Water Environ J 18(1):44–49. https://doi.org/10.1111/J.1747-6593.2004.TB00492.X

Oguchi T (2020) Lithosphere—the solid realm which supports human life. In: Himiyama Y, Satake K, Oki T (eds) Human geoscience. Springer Nature, pp 27–38. https://doi.org/10.1007/978-981-32-9224-6_3

Opoku A (2019) Biodiversity and the built environment: Implications for the sustainable development goals (SDGs). Resour Conserv Recycl 141:1–7. https://doi.org/10.1016/j.resconrec.2018.10.011

Pieterse E, Parnell S, Croese S (2017) The 2030 agenda: sustainable urbanisation and the research-policy-interface issues for the G20, pp 1–35

Rees WE (1999) The built environment and the ecosphere: a global perspective. Build Res Inf 27(4–5):206–220. https://doi.org/10.1080/096132199369336

Sakalasooriya N (2013) The principles of biophysical sustainability for sustainable development. https://www.researchgate.net/publication/283018290_

Secretariat of the Convention on Biological Diversity (2012) Cities and biodiversity outlook. In: Executive Summary. https://doi.org/10.6084/m9.figshare.99889

Seto KC, Parnell S, Elmqvist T (2013) A global outlook on urbanization. In: Elmqvist T, Goodness J, Marcotullio PJ, Parnell S, Sendstad M, Wilkinson C, Fragkias M, Guneralp B, McDonald RI, Schewenius M, Seto KC (eds) Urbanisation and global trends in biodiversity and ecosystem services. Springer, pp 1–12

Sithole A, Goredema B (2013) Building in wetlands to meet the housing demand and urban growth in Harare. Int J Humanit Soc Sci 3(8):193–201

Trotter L, Dewan A, Robinson T (1990) Effects of rapid urbanisation on the urban thermal environment in Dhaka Megacity, Bangladesh. Environ Sci 4:145–167

UNDP (2020) Rapid land-use hotspot mapping in Harare, Zimbabwe

Unganai LS (1997) Surface temperature variation over Zimbabwe between 1897 and 1993. Theoret Appl Climatol 56(1):89–101. https://doi.org/10.1007/BF00863786

Verma KA (2018) Ecological balance: an indispensable need for human SurvivaL. J Exp Zool India 21(1):407–409. www.connectjournals.com/jez

Vladimir VI (1998) The biosphere. Springer, New York

Wang C, Wang ZH, Yang J (2018) Cooling effect of urban trees on the built environment of contiguous United States. Earth's Future 6(8):1066–1081

Wang Y, Hatt JK, Tsementzi D, Rodriguez-R LM, Ruiz-Pérez CA, Weigand MR, Kizer H, Maresca G, Krishnan R, Poretsky R, Spain JC, Konstantinidis KT (2017) Quantifying the importance of the rare biosphere for microbial community response to organic pollutants in a freshwater ecosystem. Appl Environ Microbiol 83(8). https://doi.org/10.1128/AEM.03321-16

World Climate Guide (n.d.) Zimbabwe climate: average weather, temperature, precipitation, when to go. Retrieved 10 Nov 2021, from https://www.climatestotravel.com/climate/zimbabwe

World Commission on Environment and Development (1987) https://idl-bnc-idrc.dspacedirect.org/bitstream/handle/10625/152/WCED_v17_doc149.pdf?sequence=1

Geoinformatics and Land Surveying Steering the New Urban Agenda in Zimbabwe

Tarirai Pinias Masarira, Sydney Togarepi, and Juliana Useya

Abstract The world is progressively becoming interconnected and technologically dependent. Smart applications have changed approaches to everyday activities. Smart cities require data acquired through modern intelligent technology, to achieve urban sustainability. This chapter aims to explore the role of geoinformatics and surveying in advancing the New Urban Agenda (NUA) in Zimbabwe. A qualitative research design focusing mainly on the Zimbabwean case was adopted. Desktop study was mainly used, complemented by data on government and other agencies' tender notices for geoinformatics and surveying services and products. The data was then discussed according to relevant paragraphs of the NUA document. From the discussions, the geoinformatics and surveying role of establishing Zimbabwe's national coordinate system and geodetic control network provides a base for collection, integration and efficient communication of geospatial information required for urban development. The cadastral survey role is critical in developing the national cadastre and title Registration system critical in tenure security. Provision of Digital Elevation Models (DEM) and orthophotos from Unmanned Aerial Vehicles (UAVs) has enabled more detailed and accurate plans and engineering designs for the provision of housing, social amenities and infrastructure. Fit for purpose land administration principles supported by enabling legislation were recommended for dealing with informal settlements.

Keywords Geomatics · Urban sustainability · Decision-making · GeoICT · Cadastre

T. P. Masarira (✉) · S. Togarepi · J. Useya
Department of Geomatics Engineering, University of Zimbabwe, Mt. Pleasant, Box MP 167, Harare, Zimbabwe
e-mail: tmasarira@eng.uz.ac.zw

S. Togarepi
e-mail: stogarepi@gmail.com

J. Useya
e-mail: juseya@eng.uz.ac.zw

C. Chavunduka and I. Chirisa (eds.), *New Urban Agenda in Zimbabwe*,
Advances in 21st Century Human Settlements,
https://doi.org/10.1007/978-981-97-3199-2_4

1 Introduction

Urbanization is a key driver of economic development and social mobility. Globally, urban sprawl patterns in most cities have a direct relationship with technological progress. Urbanization in Africa is taking place rapidly, the rate having soared from 15% in 1960 to 40% in 2010, and being projected to reach 60% in 2050 (UNDESA 2014). This is expected to alter the profile of the region and present a huge challenge on policy makers in addressing growth and development issues. Since such growth requires good urban management greater demand will be made on the land and provision of other basic necessities. Informal settlements are already accommodating over 60% of urban Africans (Smart Africa 2017).

Escalating urban expansion demands proactive national strategies in effectively planning for existing and expected sprawl. Zimbabwe is no exception. Urban dwellers deserve and expect advanced standards of public services, housing and economic opportunities. The UN-Habitat (2016) highlighted a global trend in which cities have become central to the transformation of the global economy particularly with respect to energy consumption, water and sanitation needs, food security, tenure systems and human mobility. Consequently, governments' inability to provide these basic necessities especially in developing nations' cities has led to challenges such as informal settlement patterns and environmental degradation. To alleviate these challenges, innovative geoinformatics and surveying strategies and practitioners capable of using enhanced spatial data sensing technologies, advanced processing procedures, and related approaches have become indispensable (Tierolff 2020).

Issues related to land are often sensitive and may cause conflicts within and between communities. The complex relationship between land rights and other interests requires alignment between good policy guidelines and set developmental goals especially the New Urban Agenda (NUA). This chapter aims to review and discuss the role geoinformatics and surveying are playing in achieving the objectives of the NUA in Zimbabwe. The chapter initially gives insight into what the NUA and Geoinformatics and Surveying are. It then identifies NUA paragraphs directly related to the role of Geoinformatics and Surveying and discusses the efforts made in meeting the NUA objectives in Zimbabwe before concluding and giving recommendations on the way forward.

2 Methodology

This chapter adopted a qualitative research design focused mainly on the Zimbabwean case. Desktop study was mainly used, complemented by data on government and other agencies' tender notices for geoinformatics and surveying services and products. NUA paragraphs requiring geoinformatics and surveying services and products for their attainment were identified and organized into themes for discussion. The data obtained from the desktop studies was then discussed according to relevant paragraphs of the NUA document.

3 The New Urban Agenda

The New Urban Agenda (NUA) is a 24-page document providing global principles, policies and standards enabling sustainable urban development, focusing on transforming the way cities are constructed, managed, operated and dwelt in (Rashid 2018). It is aimed at paving the way towards making cities and human settlements more inclusive and ensuring that everyone can benefit from urbanization (Rashid 2018). Because issues related to urban and human settlements are complex, they span many sectors, agendas, policies and fields, including demographics, economics, the environment, finance, industry, politics and infrastructure (UN-Habitat 2016).

Conventional silos and fragmented approaches have hindered effective planning, implementation, monitoring and reporting of existing development agendas (Adams and Judd 2016). The NUA is expected to guide urbanization efforts for a range of actors that include national, city and regional leaders, international development financiers, the private sector, the United Nations programmes and civil society for the next 20 years (Rashid 2018). Good urbanization relies on robust urban governance structures based on supportive legal and policy frameworks, ensuring effective planning and management of urban spatial development and viable financial plans.

4 Land Surveying and Geoinformatics

Geoinformatics is the science and technology dealing with the structure and character of spatial information, its capture, its classification and qualification, its storage, processing, portrayal and dissemination, including the infrastructure necessary to secure the optimal use of this information (Raju 2006). Alternatively, it is the art, science or technology dealing with the acquisition, storage, processing, production, presentation and dissemination of geoinformation (Ehlers 2008). The related term geomatics that is commonly used these days, encompasses geoinformatics, but focuses more on surveying. Geoinformatics combines geospatial analysis and modelling, development of geospatial databases, information systems design, human–computer interaction and both wired and wireless networking technologies. At the core of geoinformatics are technologies that support the processes of acquiring, analysing and visualizing spatial data. Both geoinformatics and surveying heavily rely on the practical applications of geodesy.

Land surveying is the art, science and technology of determining or establishing the relative position of points on, above or below the earth's surface (Ghilani and Wolf 2005). It involves all methods for measuring and collecting information about the physical earth and its environment. This information is then represented as points, lines, or elevations in usable form on suitable media and scales to form plans, maps, charts, profiles and figures or stored in report tables or as digital maps with the ability to be disseminated to a wide range of clients (Ghilani and Wolf 2005).

The definition of surveying (and the name) is changing to reflect the applications of surveying techniques and the impact that modern technology has had on the more traditional aspects of the discipline. The need to mark boundaries and divide land since time immemorial has made surveying one of the world's oldest and most important professions. The discipline is critical in any physical development. Indeed, no meaningful physical development occurs without a land survey having been carried out. The ability to observe and measure aspects of the earth is critical in nation-building and has seen land surveyors, globally, being seized with the responsibility of providing accurate geospatial information for spatial planning in the urban and rural setting (Ezeomedo 2019).

5 Discussion

The identified NUA paragraphs directly related to geoinformatics and surveying were categorized into the themes: positioning in the urban space; development and enforcement of cadastres, cadastral and land registration systems; geoinformatics and surveying in urban planning; security of tenure and tenure systems; slum and informal settlement development and solutions; and the provision of social amenities and infrastructure.

5.1 Positioning in the Urban Space

Urbanization provides a tremendous opportunity and is an important tool guiding sustainable development. However, if not carefully planned and managed, urbanization can exacerbate many of the problems that it claims to solve—particularly environmental degradation, as well as increases in slums and sprawl (UN-Habitat 2016). NUA paragraph 159 highlights the important role of different kinds of information in facilitating evidence based governance and in providing a platform for building a shared knowledge base. When different attribute information of a place is given a geographical location or tag, it becomes easier to come up with a holistic overview of the different phenomena under study at any particular location. In other words, geographical position creates a base for the creation of meaningful information derived from various application domains for use in decision-making and in planning.

The importance of 'place' is realized from the fact that everything that happens occurs somewhere. Understanding the nature of 'place', where things happen, and the impact on the people and assets on that location, will facilitate better planning, risk management, and resources utilization (Geographical Information Panel 2008). Spatially enabled government is achieved when governments use 'place' as the key means of organizing their activities in addition to information, and when location and spatial information are available to citizens and businesses to encourage creativity (Enemark 2012).

The emergence and increasing popularity of 3D-Earth viewer applications such as Google-Earth have made it possible for citizens to access a wide variety of spatial data. This is especially important in developing nations where digital geospatial data is not readily available. The ability to merge 3D-Earth viewer data with geospatial data from different application domains greatly empowers governments and other stakeholders in relation to emergency response, taxation assessment, environmental monitoring and conservation, economic planning and assessment, social services planning and infrastructure planning (Enemark 2012). It also allows geospatial information on a particular place to be more efficiently organized and communicated between different users. According to Williamson et al. (2010), a spatially enabled society possesses the ability to manage information spatially as opposed to managing spatial information.

Organization of information based on geographic location requires a fundamental base on which positions can be reckoned. This base is commonly referred to as a geodetic reference system composed of a coordinate system and a datum. The datum is commonly used as a reference for elevations, and in Zimbabwe elevations are referenced to the Mean Sea Level. The coordinate reference system in Zimbabwe has as its basis as the original triangulation network composed of trigonometric beacons (Fig. 1).

Densification of the original network of beacons in Zimbabwe has resulted in around 4000 trigonometric beacons at various levels of accuracy and around 7000 Town Survey Marks being established in various parts of the country.

The positions of these beacons are expressed as spherical coordinates (i.e. latitude and longitude) on the Modified Clarke 1880 spheroid (adopted as the local best fitting spheroid for Zimbabwe). However, for large- to medium-scale mapping, engineering and cadastral surveys, projected coordinates, i.e. Eastings and Northings on the Universal Transverse Mercator (UTM) projection and the y and x on the Gauss Conformal Projection are used. The triangulation network, comprised of trigonometric beacons, is commonly referred to as a passive control network.

Most mapping and surveying projects in Zimbabwe were based on this passive geodetic network before the advent of the Global Navigation Satellite Systems (GNSS). GNSS positioning commonly relies on a global best fitting ellipsoid known as the World Geodetic System of 1984 (WGS84). An increase in the procurement and use of GNSS equipment by geoinformatics and surveying practitioners combined with the need for increased turnaround times of surveys has seen the Government of Zimbabwe (the Surveyor General) establishing a network of Continuously Operating Reference Stations (CORS). The CORS network is comprised of GNSS receivers that operate continuously throughout the day transmitting correction information to GNSS receivers in the field via the internet. This network is commonly referred to as an active survey control network.

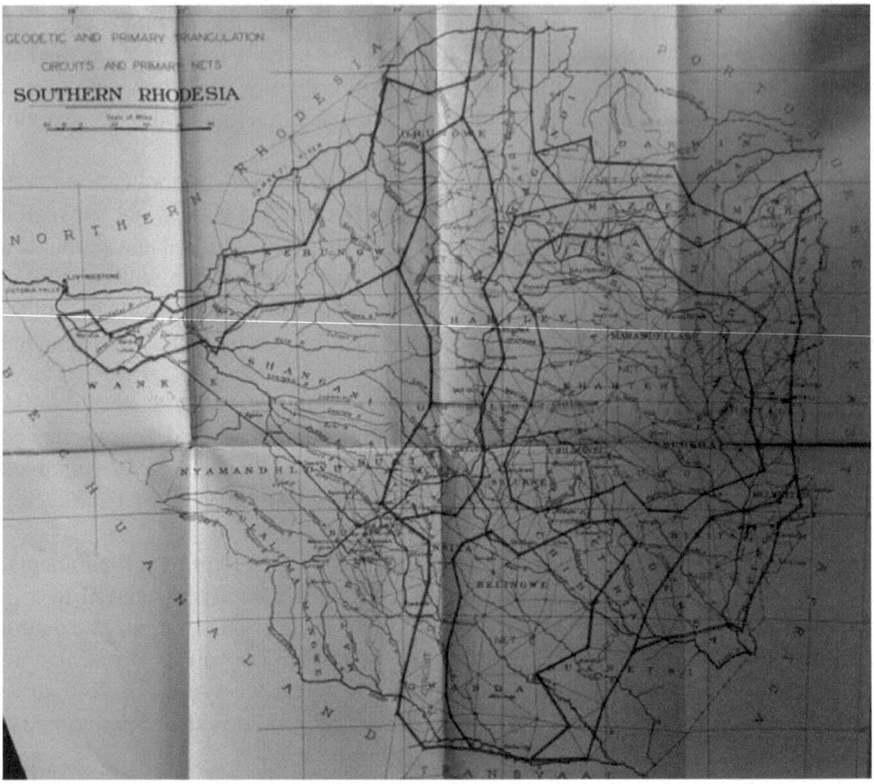

Fig. 1 The primary geodetic network of Zimbabwe as comprised of trigonometric beacons. *Source* The Surveyor General Zimbabwe

Currently, the passive and active control networks complement each other and have significantly improved the efficiency of carrying out surveys. Although the two networks use different coordinate referencing systems, knowledge of the relationships between the different ellipsoids used allows calculation of parameters that enable accurate transformation of data from one system to the other. This enables data from different sources to be accurately merged and utilized.

The control points forming both the passive and active control networks are established using the branch of land surveying called geodesy. Geodetic surveying is concerned with survey measurements for the determination of the size and shape of the earth including defining the size, position, shape and elevation of any objects in space, on, above or below the surface of the earth using surveying instruments (Ghilani and Wolf 2005). Planning, designing and implementation of cadastral, engineering and construction projects in Zimbabwe are preceded and dependent on the establishment and maintenance of existing geodetic networks.

5.2 *The Cadastre and Cadastral System*

The NUA paragraphs 104, 134 and 156 highlight the importance of encouraging the development, establishment and use of digital platforms and tools, including geospatial information systems, to improve long-term integrated urban and territorial planning and design, land administration and management, and access to urban and metropolitan services. Zimbabwe's information and communications technology policy and e-government strategies including capacity-development programmes aim to make information and communications technologies accessible to the public. This is envisaged to lead to the development and exercise of civic responsibility, broadening of participation and fostering of responsible governance, as well as increasing efficiency.

The development of cadastres (as per NUA paragraphs 104, 134 and 156) results from survey measurements made during the survey processes of land subdivision and consolidation. These processes also involve the creation and demarcation of subsidiary rights to land such as servitudes and leases in compliance with the legal requirements enforced by accountable institutions that deal with land registration and governance, and property registration (NUA paragraph 104). Survey processes done to achieve the above are defined as cadastral surveys. In Zimbabwe cadastral surveys can only be legally undertaken by land surveyors registered by the Council of Land Surveyors (CLS) and result in large scale survey plans or 'diagrams'.

Basically, in Zimbabwe, the legislation regulating the development of basic land inventory information, such as cadastres, are the Land Survey Act, the Land Surveyors Act, the Regional Town and Country Planning Act, the Deeds Registries Act together with their accompanying regulations, administered and enforced by the Surveyor-General, CLS, the Director of Physical Planning and the Registrar of Deeds, respectively. Zimbabwe's cadastral system comprises two components, namely, a legal and technical cadastre. The legal cadastre deals with the process of registering land and interests in land while the technical cadastre is concerned with the technical process of property boundary demarcations and plan production.

Geoinformatics and surveying play an important role in the cadastral system in that they provide the evidence required for proving the ownership of a property in cases of dispute or land settlements. An accurate survey solves a lot of misunderstanding when it comes to property ownership (Ezeomedo 2019). Surveyors stand as witnesses or advisers, in the event of land litigation. If an urban area grows to the extent that land encroachment occurs, the surveyor usually is called upon as the expert in the land case (Brinker and Minnick 1995).

The term cadastre is generally applied to a continuous (seamless) map of surveyed properties and associated attributes (that describe the nature of interests, ownership or control of interests, and often the value of land parcels and its improvements). It forms the technical core of a spatially enabled government. The International Federation of Surveyors defines a cadastre as a parcel-based and up-to-date land information system containing a record of interests in land (e.g. rights, restrictions and responsibilities). The cadastre may be established for fiscal purposes (valuation and taxation), legal

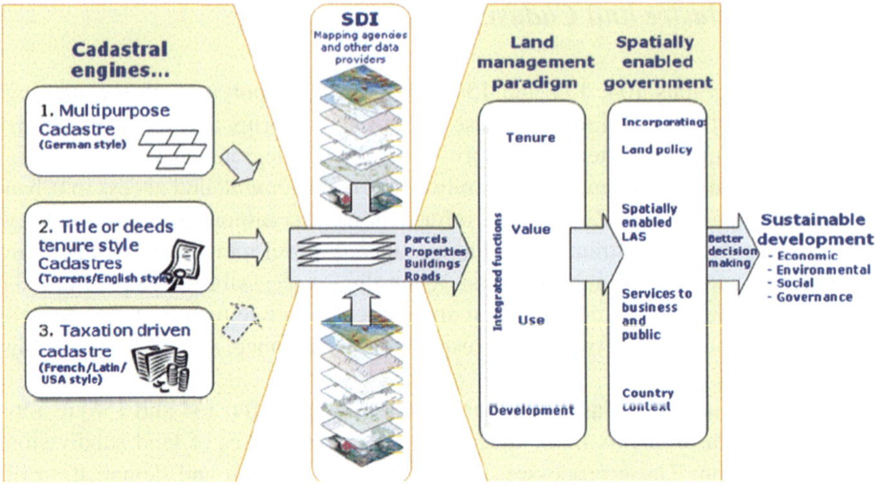

Fig. 2 Cadastre as an engine of LAS—the "butterfly" diagram. *Source* Williamson et al. (2010)

purposes (conveyancing), to assist in the management of land and land-use control, and enables sustainable development and environmental improvement.

The cadastre acts as an engine of Land Administration System (LAS) as shown in Fig. 2 and Zimbabwe employs the title deed style cadastre. In Zimbabwe's case, the cadastre is composed of the cadastral map (housed in the Surveyor-General's office) and the land register (housed in the Deeds Registry). These two organizations fall under two different ministries, and this has led to parallel development of the two cadastral system components resulting in digital information systems that do not communicate efficiently. This obviously has negative effects on the implementation of the NUA in Zimbabwe.

Figure 2 highlights the usefulness of the large scale cadastral maps as tools that represent the scale of land use and how people are connected to their land. Wherever the cadastre sits in a national land administration system, it should assist the functions of land tenure, value, use and development. In this way, the cadastral system becomes the core technical engine delivering the capacity to control and manage land through the four land administration functions (Enemark 2012). It supports business processes of tenure and value, depending on how the cadastre is locally built and identifies the legal rights, the units that form the commodities and the economy related to property. These cadastres are much more than a layer of information in National Spatial Data Infrastructure (SDI) (Enemark 2012).

Figure 2 can be likened to a virtual butterfly where one wing represents the cadastral processes, and the other the outcome of using the processes to implement the land management paradigm (Enemark 2012). Once the cadastral data are integrated within the SDI, the full multipurpose benefit of the LAS can be achieved (Enemark 2012). The body of the butterfly is the SDI, with the core cadastral data sets acting as the connecting mechanism thus taking the importance of cadastral information beyond the land administration framework by enlarging its capacity to service other essential functions of government, including emergency management, economic management, effective administration, community services and many more (Enemark 2012).

Additionally, a cadastre should have the capacity of helping users identify a parcel of land both on the ground and in the system in ways they can relate to, typically an address plus a unique systematically generated identifier (Enemark 2012). According to Enemark (2012), the core cadastral information of parcels, properties, buildings, vested roads and real rights in land, thus becomes the core of SDI information. In Zimbabwe, the SDI is under the purview of the Surveyor General's office as covered under Sect. 5.2 of the Land Survey Act. However, despite a number of research studies having been done, no visible action regarding implementation has been seen on the ground.

The Zimbabwean cadastre, comprising surveyed land units and in some cases buildings, covers most of the urban areas. However, most of this information is in hard copy format stored and maintained by the Surveyor-General. Some town and city councils notably Harare, Bulawayo, Kwekwe, Victoria Falls, Chinhoyi including private companies and individuals have created digital cadastres of properties in their jurisdictions. On the other hand, the Surveyor General is also converting cadastral data under its custodianship into digital format under the EU and UNDP support to the Ministry of Lands and Rural Resettlement 2014–2016 and a government tender for the establishment of an e-cadastre was floated in March 2022. Unfortunately, there is little evidence of direct cooperation and sharing of this digital information between the Surveyor General's office and the other players.

5.3 Surveying and Geoinformatics in Planning

Effective planning requires topographic data at suitable scales, cadastral information, various urban plans and engineering charts of the project area (Shwarts and Felus 2006). It is underpinned by adequate input from allied professions such as geoinformatics and surveying, engineering and the prevailing legal systems, among other local facts. These facts include the natural environment, relief and topography, information on the existing statutory conditions of land use, i.e. cadastral information (Shwarts and Felus 2006). This is aptly captured in Fig. 3 showing the surveyor's role in urban development.

Plans produced not only for housing development, but also for layouts of commercial and industrial areas, route designs, site selection as well as land-use planning, require an accurate topographical map as a base. Topographical maps are normally produced from topographical surveys, undertaken primarily to determine the positions of objects (both natural and man-made features) on the earth's surface in both plan and height.

According to Socrates, (4th century, BC) "*Man must rise above the earth-to the top of the atmosphere and beyond; for only thus will he fully understand the world in which he lives*". This statement has implications for planning since design plans

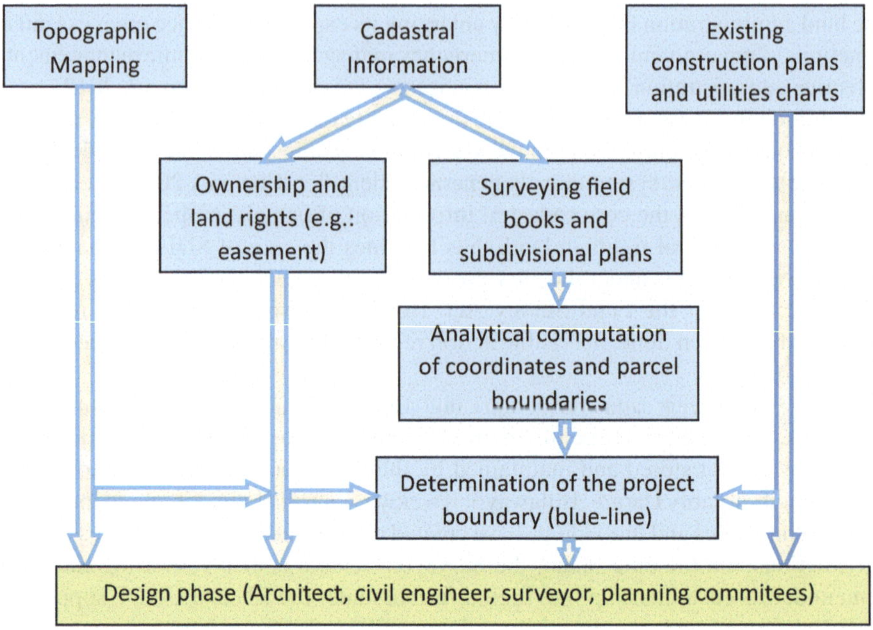

Fig. 3 Urban development workflow. Surveyor tasks are shown by blocks in light blue. *Source* Shwarts and Felus (2006)

of an area are influenced by what is happening in the area abutting the site, making a holistic view of the site and its surroundings critical. It also has implications on geoinformatics and surveying in that data collection efforts should lead to products that enable users to realize a holistic view of the planning area. Traditionally, in Zimbabwe, localized large-scale topographical maps were produced using ground based methods that utilized optical and distance measurement equipment, while medium- to small-scale maps were produced using aerial photogrammetric methods.

Some geoinformatics and surveying practitioners have embraced and moved in pace with advances in science and technology and are now undertaking topographic surveys using a combination of GNSS equipment, aerial/drone photogrammetric methods and satellite imagery. In addition to topographic maps showing the topography, drainage patterns, settlements types and patterns, different types of infrastructure, vegetation and land cover types among other features, practitioners are also producing 3D digital surface models (DSM) and digital elevation models (DEM) of an area, all of which can be used for urban geospatial analysis and decision-making. The dilemma, however, is that the intended beneficiaries have been very slow in embracing these products and using them for urban planning, infrastructural designs and development control purposes in Zimbabwe.

Topographic mapping may be done to meet specified project requirements depending on intended planning outcomes and also the characteristics of the project area (Shwarts and Felus 2006). In carrying out the topographical survey, a balance should be struck between the amount of detail, the accuracy and cost. This calls upon the surveyor being able to choose appropriate survey methods and equipment to meet specified accuracy requirements of the given map scale as well as to minimize costs (Table 1).

Surveying and geoinformatics practitioners' skills should also include planning and estimating new engineering projects, marking and delimiting boundaries and setting out engineering works (McKay 2019). The surveyor's project management skills should facilitate the integration of various data sets and skills in an interdisciplinary and inter-professional way. This involves partnering a diverse range of stakeholders from the international aid community, professional organizations, the academic community and key decision-makers (FIG 2008). The ability to facilitate partnerships between the people who implement land planning and those with technology expertise (FIG 2008) has seen some geoinformatics and surveying practitioners in Zimbabwe becoming agents for global software and equipment manufactures. This has afforded other professionals in Zimbabwe easier access to software and equipment.

More than ever before, there has been an increase in Requests for Proposals or Tenders for the production of topographic maps of pieces of land of different sizes in Zimbabwe. These calls have originated from different administrative levels of government for the production of basemaps mainly for preparation of Master Plans, e.g. for Tugwi-Mukosi, the New City Master plan and the Sakubva Master plan for Mutare. Topographical basemaps have also been requested for the Gwayi-Shangani catchment development project as well as for the Kunzvi Dam pipeline route development. This has been due to the fact that the national mapping series (the

Table 1 Topographic maps used in the planning process according to terrain conditions, *Source* Shwarts and Felus (2006)

Type of spatial database	Type of area	Basic layers	Output map scale	Contour interval	Surveying method
Large scale	Densely populated urban area	Structures, transportations, hydrology (curb and gutter), land use, utilities (manholes), vegetation, (single trees and bushes)	1:500	0.5 m	Land surveying (total station) and photogrammetry
Intermediate scale	Moderately built-up and developed area	Structures, transportations, hydrology, land use, utilities, vegetation	1:1000 1:1250	1 m	Photogrammetric mapping using low altitude aerial photographs
Small scale	Rural area	Structures, transportations, hydrology, land use, major utilities	1:2500	2.5 m	Photogrammetric mapping using low altitude aerial photographs
Very small scale	Open area (agriculture)		1:5000	5 m	

1:50,000 scale for large areas and the 1:5000 scale for smaller areas) are largely out of date. Topographical maps can now be done easily using drone technology to produce accurate and up to date digital map data. Other popular modern alternatives include high resolution satellite imagery as well as Volunteered Geographic Information (Keenja et al. 2012; Moreri et al. 2018).

Changes in the Information Communication Technology (ICT) industry are occurring very fast, and the ability to utilize ICT has become of strategic importance. Geographic Information Systems (GIS) topographic mapping has become a powerful vehicle in developing land policies and addressing the NUA focal points on promoting the development of urban spatial frameworks and urban renewal to prevent urban sprawl and marginalization. Use of GIS in urban areas facilitates successful planning work (FIG 2008) and implementation of the NUA. In Zimbabwe, GIS technology is neither readily available nor accessible, and the various methods used to access ICT have only recently started being incorporated into the planning sector. Different professionals in Zimbabwe, though, have started appreciating the importance of map-based services as crucial in the delivery of the sustainability agenda. This has been encouraged by the introduction of geospatial-based courses at high school level and in higher and tertiary institutions' programmes across a variety of professional domains.

5.4 Security of Tenure and Tenure Systems

Through the NUA paragraphs 14(b) and 35, governments committed themselves to promote: increased security of tenure for all; recognition of the plurality of tenure types; development of fit-for-purpose and age, gender and environment responsive solutions within the continuum of land and property rights; the prevention of land speculation; and managing urban shrinking where possible.

Security of tenure is the right of all individuals and groups to protection by the state against forced evictions, and according to Payne et al. (2012), describes an agreement between an individual or group to land and residential property, which is governed and regulated by a legal and administrative framework. Security of tenure derives from the fact that the right of access to and use of the land and property is underwritten by a known set of rules, and that this right is justifiable. A housing structure should be accompanied by security of tenure, i.e. members of a household should have legal, social or other protection against arbitrary unlawful eviction, harassment as well as any other threats (UN-Habitat 2018b). The housing structure should have statutory or customary law or informal or hybrid arrangements that safeguard the house members against forced evictions. In most developing nations' urban settlements, slum dwellers lack security of tenure for their housing units (UN-Habitat 2018b).

The NUA stresses that no one should be left behind concerning land rights and this has been enshrined in *Zimbabwe's Vision2030*. Governments committed themselves to recognize, respect and safeguard all legitimate tenure rights along the continuum of

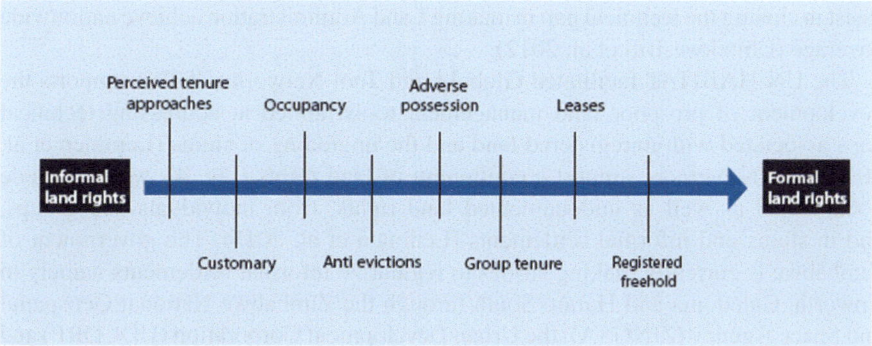

Fig. 4 Continuum of land rights, *Source* UN-Habitat (2008)

rights (Fig. 4), including customary and informal rights, and to promote and facilitate their enjoyment. Urban land in Zimbabwe is predominantly under freehold and leasehold tenure. These tenure types place a lot of emphasis on accuracy (Bennett et al. 2012) and thus much interest from surveyors. However, survey practitioners recently started being involved in discussions and research on the national land policy and the establishment of a cadastre covering all the other types of tenure including mining and mineral rights. This may be attributed to a move from traditional surveying curricula in tertiary institutions towards land administration and management.

Legal or formal systems by nature do not serve millions of people whose tenures are more social rather than legal. The way land rights are currently structured under freehold and registered leasehold (including conventional cadastral and land registration systems) fall short in delivering security of tenure to the low income groups (especially in informal settlements) thus innovative approaches that broaden access to land are required (UN-Habitat 2018a). This should include a 'scaling up approach' from informal to more formalized land rights, but this does not necessarily mean all societies developing into freehold tenure systems. Figure 4 shows a continuum of land rights where stages in the process can be formalized, with registered freeholds offering a stronger protection, than at earlier stages (UN-Habitat 2008).

Most developing countries have over 70 per cent of the land existing outside the land register (Lemmen et al. 2015), resulting in enormous problems for example in cities, where people living in slums have no security of tenure. Security of tenure for people in these areas is based on forms of tenure different from individual freehold. Most off register rights and claims are based on social tenure, for which the Social Tenure Domain Model (STDM) was developed to accommodate. This is a pro-poor land information management system aimed at supporting land administration for the poor in urban and rural areas. Ideally this should be linked to the national cadastral system in order for all land information to be integrated for sound Land Administration practices (Lemmen et al. 2015). The STDM was piloted in Epworth Zimbabwe by Dialogue on Shelter and proved that it could be customized and applied in the Zimbabwean context to address the issue of informal settlements as well as assist in closing the technical gap in making Land Administration achieve nationwide coverage (Chitekwe-Biti et al. 2012).

The UN-HABITAT facilitated Global Land Tool Network (GLTN) supports the development of pro-poor land management tools, aimed at addressing technical gaps associated with unregistered land and the upgrading of slums (Lemmen et al. 2015). GLTN partners support a continuum of land rights (Fig. 4), which include documented as well as undocumented land rights, from individuals and groups, and in slums and informal settlements (Lemmen et al. 2015). The government of Zimbabwe is currently making strides to regularize informal settlements namely in Epworth, Caledonia and Harare South through the Zimbabwe National Geospatial and Space Agency (ZINGSA), the Urban Development Corporation (UDCORP) and some development partners like Dialogue on Shelter Zimbabwe.

Other approaches to achieving security of tenure for all include the Framework for Effective Land Administration (FELA) and the Fit For Purpose Land Administration (FFPLA). According to Metaferia et al. (2023) FELA aims to combat worldwide tenure insecurity through rapid and sustainable land administration. On the other hand FFPLA principles aim at achieving complete coverage of all land in terms of identification and documentation with improvements taking place over time in line with legal requirements, improved accuracy and institutional processes (Metaferia et al. 2023). The Namibian flexible tenure system supported by the Flexible Land Tenure Act (No. 4 of 2012) appears to be an example of the FFPLA and is increasing security of tenure in the informal settlement parts of urban areas in Namibia.

A complete coverage of all land by Land Administration Systems will not only allow for registration of formal rights and recordation of informal and customary rights, but also for managing the value and use of land and land development plans (Enemark 2012). According to Enemark (2012) extendable and flexible models such as STDM can facilitate the inclusion of all the land and people within the four land administration functions outlined in Fig. 5.

5.5 Slum and Informal Settlement Development and Solutions

Rapid growth often results in vulnerability to climate and disaster risks in urban areas. Effective planning and administrative systems are critical to avoid continued developments on vulnerable and disaster-prone areas and to mitigate pressure on land. Informal settlers often settle on vulnerable and disaster-prone areas, leading to policy challenges related to relocation, compensation and the use of informal settlements for political bargaining. In Zimbabwe, addressing informality is made difficult, not only due to limited financial resources and technical know-how but also the existing legislation and land policies.

The NUA paragraphs 107—110 focus on combating homelessness, promoting the upgrading, prevention and integration of slums and informal settlements through

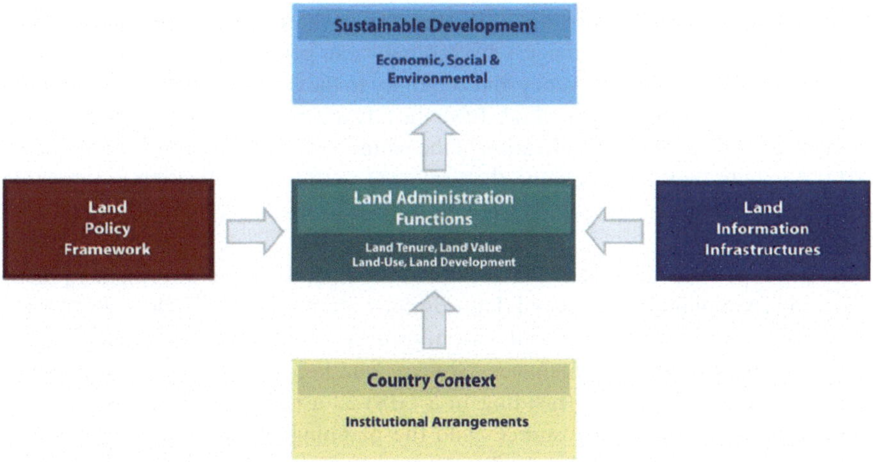

Fig. 5 Land management paradigm, *Source* Enemark (2012)

appropriate policies and strategies targeting the promotion of security of tenure and its regularization. Signatories committed themselves to support efforts to develop monitoring systems aimed at reducing the number and improving the living conditions of slum and informal settlement dwellers. Discussions also focused on the need for alternative forms of tenure to address the evolving needs of communities, to facilitate housing provision and to prevent segregation and arbitrary forced evictions and displacements. In an effort to address these slum predicaments, Mahabir and Crooks (2018) proposed a five-step approach comprising the following: define, data, map, analyse and model (DDMAM). The efforts are to enhance provision of better infrastructure and services by availing logical and scientific information for decision- and policy-making.

In Zimbabwe, efforts to regularize the informal settlements of Epworth and Caledonia involved participation of residents in replanning leading to surveys of the planned portions. The surveys were however fraught with problems largely because layout plans were not based on accurate and up-to-date topographical base maps. Drone mapping has proved critical in producing accurate topographic base maps derived from orthophotos and digital elevation models (DEM) in relatively short times. Accurately georeferenced orthophotos present enormous possibilities for authorities to come up with innovative forms of tenure to address the plight (security of tenure) of intended beneficiaries.

5.6 Provision of Social Amenities and Infrastructure

For precise positioning of engineering works an accurate and reliable geoidal model is required. Initiatives to develop a local geoidal model for Zimbabwe are underway, which will be useful in sustainable infrastructural development for smart cities. The use of airborne and satellite imagery by Zimbabwean geoinformatics and surveying practitioners has brought efficiency and economy to the development of base data for designing and planning. The methods have the advantage of producing products that enable three-dimensional visualization of the natural and built environment (Ndukwe 2001). As such, they have become the preferred option for collecting topographic data over vast, complex, irregularly shaped and precarious sites (Birriel and González 2016).

Technological advancements have created a demand for geoinformatics practitioners in urban planning as spatial data administrators seized with the role of organizing different raster and vector files from project inception, as-built mapping and ultimately infrastructure maintenance (Shwarts and Felus 2006). Indeed the role of accurate spatial information (residing in a GIS) as a rich source of basic data in the management of natural resources and the planning of future developments is gaining wider recognition by urban planners and engineers (Jensen 2007; Ezeomedo 2019; Sliuzas and Brussel 2000). In these instances, geoinformatics and surveying practitioners are usually always behind the scenes updating data on the built environment and correcting errors in older survey documents caused by obsolete methods of data collection. However, Chukwu (2015) notes that these practitioners remain invisible to the average citizen until such a time when their service is required, yet they are indispensable, and crucial in the country's institutional and infrastructural development.

In Zimbabwe, the recognition of the geoinformatics and surveying role has been brought to the fore particularly by the ongoing National and Emergence Road Rehabilitation programmes. This has led to an increase in the rate of employment or engagement of these practitioners major engineering firms, municipalities and government departments in various engineering works as the need for geospatial information increased. Geospatial information especially in the form of orthophotos and DEMs has the potential of assisting planners and engineers to plan effectively for the provision of basic social amenities and infrastructure. At the localized level, DEM data can effectively inform architects to develop suitable and cost effective housing solutions for urban settlers –in both informal and formal settlements. DEMs, orthophotos and topographical basemaps derived from drone imagery were indispensable in the development of Zimbabwe's New City and Tugwi-Mukosi master plans. The products presented many design options to planners and architects and facilitated a more holistic developmental process with possibilities of enabling simulations as well as model development. The demand for these products can only grow as the Zimbabwean Ministry of Local Government currently demands 3D visualizations of proposed plans before approval of housing development layout plans.

6 Conclusion and Recommendations

The chapter reviewed and explored the role of geoinformatics and surveying in the realization of the objectives of the NUA by identifying relevant NUA paragraphs that were directly impacted by current geoinformatics and surveying practices, products and services. The critical role of having a common base or platform (the national geodetic reference system) that enables spatial data to be accurately mapped and integrated and major stakeholders to collaborate was highlighted. This National Geodetic System is indispensable in boundary delimitation and demarcation leading to the creation of the national cadastre, land and land rights registration leading to enhancement of security of tenure.

The adoption of modern GeoICT to collect and manipulate geospatial information has brought efficiency to planning for smart and sustainable urban development. Products such as Orthophotos, 3D DEMs have presented greater opportunities for visualization, modeling and simulations for more resilient and robust designs and plans for route, drainage, building and construction design and planning. Correctly georeferenced orthophotos used simultaneously with digital cadastres assist in development control and identification irregular and informal settlements, thereby presenting options for their regularization.

Dealing with informal settlements still presents a challenge in Zimbabwe not only due to limited resources but mostly due to the existing legislation. Informal settlements on urban state and municipal land have a chance of being addressed using innovative approaches such as the STDM, FFPLA, FELA and the DDMAM among others unlike on land under Freehold. However, there is need for enactment of supportive legislation that also supports registration of other forms tenure apart from Freehold and Leasehold.

Despite the availability of accurate geospatial information, it is important for stakeholder professions to develop the capacity to utilize the information for successful implementation of the NUA. Lastly it is critical for the national authorities controlling the production and distribution of geospatial information to efficiently cooperate to enable access to this critical resource by the user community.

References

Adams B, Judd K (2016) 2030 agenda and the SDGs: indicator framework, monitoring and reporting. Agenda 10:18

Bennett R, van der Molen P, Zevenbergen J (2012) Fitted, green, and volunteered: legal and survey complexities of. Geomatica 66(3):181–193

Birriel P, González R (2016) PHOTOGRAMMETRIC SURVEYS OF UAVs as tools for urban planning in Uruguay. GIM Int 6–7. Available at: https://www.gim-international.com/content/article/uav-as-a-tool-for-urban-planning

Brinker RC, Minnick R (1995) The surveying handbook. Springer Science & Business Media

Chitekwe-Biti B et al (2012) Developing an informal settlement upgrading protocol in Zimbabwe—the Epworth story. Environ Urban 24(1):131–148. https://doi.org/10.1177/0956247812437138

Chukwu L (2015) Raising the bar in surveying for national development. The Guardian. Available at: https://guardian.ng/features/focus/raising-the-bar-in-surveying-for-national-development/. Accessed 15 Sept 2021

Ehlers M (2008) Geoinformatics and digital earth initiatives: a German perspective. Int J Digital Earth 1(1):17–30

Enemark S (2012) Sustainable land governance. FIG Working Week, Rome, Italy, pp 6–10

Ezeomedo I (2019) Role of land surveyors and geospatial engineers in the built environment. 1 Jan 2019 (March), pp 102–117. Available at: https://www.researchgate.net/publication/339612853_ROLE_OF_LAND_SURVEYORS_AND_GEOSPATIAL_ENGINEERS_IN_THE_BUILT_ENVIRONMENT

FIG (2008) Informal settlements: the road towards more sustainable places. Copenhagen, Denmark. Available at: http://www.fig.net/resources/publications/figpub/pub42/figpub42_A4_web.pdf

Geographical Information Panel (2008) Place matters: the location strategy for the United Kingdom. Available at: file:///C:/Users/User/Downloads/uk-location-strategy_10.pdf

Ghilani C, Wolf P (2005) Elementary surveying: an introduction to geomatics. Preentice Hall, Upper Saddle River

Jensen JR (2007) Remote sensing of the environment: an Earth resource perspective, 2nd edn. Prentice Hall, Upper Saddle River, NJ

Keenja E, De-Vries W, Bennett R (2012) Crowd sourcing for land administration: perceptions within Netherlands Kadaster. In: FIG Working Week 2012: knowing to manage the territory, protect the environment, evaluate the cultural heritage (May 2012), pp 6–10

Lemmen C, van Oosterom P, Bennett R (2015) The land administration domain model. Land Use Policy 49:535–545. https://doi.org/10.1016/j.landusepol.2015.01.014

Mahabir R, Crooks A (2018) Mapping and monitoring of slums in 5 steps|GIM international. GIM Int 1–2. Available at: https://www.gim-international.com/content/article/mapping-and-monitoring-of-slums-in-5-steps

McKay DR (2019) What does a surveyor do? The balance careers. Available at: https://www.thebalancecareers.com/surveyor-job-duties-salary-requirements-4146978. Accessed 15 Sept 2021

Metaferia MT et al (2023) Fit-for-purpose land administration and the framework contemporary experiences. Land 12(58). https://doi.org/10.3390/land12010058.

Moreri KK, Fairbairn D, James P (2018) Volunteered geographic information quality assessment using trust and reputation modelling in land administration systems in developing countries. Int J Geogr Inf Sci 32(5):931–959. https://doi.org/10.1080/13658816.2017.1409353

Ndukwe NK (2001) Digital technology in surveying and mapping: principles, applications and legislative issues. Rhyce Kerex

Payne BG, Durand-lasserve A, Payne G (2012) Holding on: security of tenure—types, policies, practices and challenges. In: Special rapporteur on adequate housing as a component of the right to an adequate standard of living, and on the right to non-discrimination (October), pp 1–78. Available at: http://www.ohchr.org/Documents/Issues/Housing/SecurityTenure/Payne-Durand-Lasserve-BackgroundPaper-JAN2013.pdf

Raju PLN (2006) Fundamentals of geographical information system. In: Satellite remote sensing and GIS applications in agricultural meteorology, p 103

Rashid AA (2018) Action framework for implementation of the new urban agenda (AFINUA) towards sustainable urban development and community wellbeing. In: The fourth industrial revolution impact to the future of sustainable urban development in Malaysia. URBANICE Malaysia, pp 1–25. Available at: https://www.academia.edu/36209789/ACTION_FRAMEWORK_FOR_IMPLEMENTATION_OF_THE_NEW_URBAN_AGENDA_AFINUA_TOWARDS_SUSTAINABLE_URBAN_DEVELOPMENT_AND_COMMUNITY_WELLBEING

Shwarts O, Felus M (2006) The surveyor's role in the development of an urban construction plan in Israel. In: Shaping the change XXIII Fig congress, Munich, Germany: FIG, pp 1–8. Available at: https://www.fig.net/resources/proceedings/fig_proceedings/fig2006/papers/ps07/ps07_12_shwarts_felus_0359.pdf

Sliuzas R, Brussel M (2000) Usability of large-scale topographic data for urban planning and engineering applications. Int Archives Photogrammetry Remote Sens 33(B4/3; PART 4), pp 1003–1010
Smart Africa (2017) Smart sustainable cities: A blueprint for Africa. The Smart Cities Blueprint
Tierolff F (2020) Smart surveyors: developments and trends from the FIG working week. GIM Int
UN-Habitat (2008) Secure land rights for all, Nairobi, United Nations Human Settlements Programme
UN-Habitat (2016) World Cities Report 2016: Urbanization and Development—Emerging Futures. https://unhabitat.org/sites/default/files/download-manager-files/WCR-2016-WEB.pdf
UN-Habitat (2018a) SDG Indicator 1.4.2 Training Module: Secure Tenure Rights To Land. United Nations Human Settlement Programe, Nairobi, pp 1–17. Available at: https://unhabitat.org/sites/default/files/2020/07/indicator_1.4.2_training_module_secure_tenure_rights_to_land.pdf
UN-Habitat (2018b) SDG indicator 11.11 training module: adequate housing and slum upgrading. Narobi, Kenya, pp 1–31. Available at: https://unhabitat.org/sites/default/files/2020/06/indicator_11.1.1_training_module_adequate_housing_and_slum_upgrading.pdf
UNDESA (United Nations Department of Economic and Social Affairs) (2014) World urbanization prospects: the 2014 revision. Demographic research. UNDESA New York
Williamson I et al (2010) *L*and administration for sustainable development. Citeseer

Spatial Planning Steering the New Urban Agenda in Zimbabwe

Charles Chavunduka, Queen Linda Chinozvina, and Kadmiel Wekwete

Abstract A spatial planning system that effectively fits in the built environment disciplines is imperative to the development of sustainable human settlements. However, Zimbabwe's current spatial planning system has not defined a niche in the built environment disciplines in a way that optimises its role. To highlight the importance of shaping the spatial planning system for sustainable human settlements, this chapter explores theoretical and practical dimensions that undergird its utility in the Zimbabwe context. The methodological approach consists of a literature review and key informant interviews. The functionality of the spatial planning system in the built environment is analysed including changes that may be considered to enable it to work more efficiently in the future. Although all spatial planning systems have one goal—organising the use of land to encourage development and facilitate the supply of services necessary for economic and social development—the results reveal that the system is hampered by a development control focus, lack of urbanisation policy, corruption as well as an unregulated profession. The way forward will require the adoption of a policy mix that includes strategic spatial planning, a national urban policy, and the establishment by the national leadership of safeguards against improper use of spatial planning powers.

Keywords Spatial · Planning · Sustainable · Built environment · Zimbabwe

C. Chavunduka (✉)
Faculty of Engineering and the Built Environment, University of Zimbabwe, Mt. Pleasant, Box MP 167, Harare, Zimbabwe
e-mail: cchavunduka@eng.uz.ac.zw

Q. L. Chinozvina
Department of Architecture and Real Estate, University of Zimbabwe, Mt. Pleasant, Box MP 167, Harare, Zimbabwe

K. Wekwete
Midlands State University, Gweru, Zimbabwe

C. Chavunduka and I. Chirisa (eds.), *New Urban Agenda in Zimbabwe*,
Advances in 21st Century Human Settlements,
https://doi.org/10.1007/978-981-97-3199-2_5

1 Introduction

Sub-Saharan Africa is experiencing rapid urbanisation as large numbers of people migrate from rural areas to existing and new urban centres. It is estimated that by 2050, over 50% of the population in sub-Saharan Africa will live and work in urban centres (Oppong 2006). This is a major challenge in the context of the current global recession that began in 2008 and has accelerated economic restructuring and rapid growth of unemployment in all parts of the world. One important effect of these economic processes has been the rapid growth in the informal economy in all urban centres, but particularly in developing countries (Machakaire 2015; UN-Habitat 2009). The broader context of African cities goes beyond rapid urbanisation and informality to include poverty, inequality, safety and climate change—induced environmental crisis.

It is against this backdrop of emerging urban challenges that the New Urban Agenda (NUA) was adopted at the United Nations Conference on Housing and Sustainable Urban Development (Habitat III) in Quito, Ecuador on 20 October, 2016. The NUA represents a globally shared vision for a better and more sustainable future. It is aligned with Sustainable Development Goal (SDG) 11 on making cities and human settlements inclusive, safe, resilient and sustainable. But all the 17 SDGs contribute to a sustainable future. The NUA reflects the importance of harnessing the urbanisation process for development globally and in Africa where urbanisation is proceeding faster than in any other region of the world. If planned and well managed, urbanisation can be a powerful tool for sustainable development for both developing and developed countries (Caprotti et al. 2017). Many African cities are experiencing rapid growth of the informal sector which in most cases evades formal planning processes (Wekwete 2018). Cities have become unsafe due to lack of clean and safe water and sanitation services, and increasing crime rates. Inadequate housing services and poor security have become part of people's daily lives. Zimbabwean cities have not been spared these troubles because inequality and poverty have been on the rise in both urban and rural areas (Kanyenze 2021).

The New Urban Agenda assigns a central role to spatial planning at national, regional and local levels for implementation of SDG 11. This role is promoted through emphasis on regulation, along with concepts on territorial and regional balance (UN-Habitat 2016; Watson 2016). The NUA makes a case for meaningful urban interventions and this entails need for clarity concerning the role of spatial planning in the agenda's enablement. Spatial planning systems in sub-Saharan Africa and Zimbabwe in particular tend to be weak, under-resourced and highly centralised; and may not have the ability to reverse trends of inequality and exclusion as expected by the NUA (Watson 2016). In Zimbabwe like in many developing countries, most growth is taking place at the city margins through informal development by developers and cooperatives and often has little to do with the efforts of planners. Those living in informal settlements find their own ways to access housing, economic opportunities and social amenities, often rendered illegal in official policies (Caprotti et al. 2017; Mehaffy and Haas 2020; Watson 2016). In a challenging context, this chapter examines how spatial planning has been promoting the NUA in Zimbabwe. It further considers how it can be shaped to more effectively enable the NUA.

2 Methodology

The chapter is largely based on existing literature and data obtained from review reports and key informant interviews. Since the inception of the NUA in 2016, Zimbabwe has submitted progress reports on the implementation of SDGs to the United Nations High-Level Political Forum in 2017 and 2021. In 2017 the country completed both a voluntary national and sub-national review of SDG implementation. In 2021, only the voluntary national review was done. Key informant interviews were important for purposes of obtaining first-hand information from experts in spatial planning based on their knowledge and experience. Interviews were conducted with two senior planners in the Government Department of Spatial Planning and Development and Harare City Council respectively. The interviews were also used for purposes of validating findings from the review reports. Thematic analysis was used to derive meaning from the data. Key themes and concepts were derived from the literature mapping the content to be reviewed and analysed. The themes were related to spatiality mainly focussing on the influence of spatial planning activities on the achievement of the NUA.

3 Conceptual Framework

There are multiple ways of looking at spatial planning in general and specific to local institutional, development or topographic contexts. In most countries, spatial planning is concerned with identifying long- or medium-term objectives and strategies for territories, and dealing with land use including physical development as a distinct sector of government activity and coordination of sectoral policies such as transport, environment and agriculture (Stead and Nadin 2008). The policies should have an impact and most importantly be capable of delivering programmes that influence both the nature and function of places.

Spatial planning is integrative in nature and aims to ensure integrated land use planning to address spatial imbalances of the past, assist in the creation of enabling environments, improve economic rural–urban linkages through development corridors, ensure compact urban areas, support active public participation, ensure sustainable environments with a sense of place and viable local communities. It is concerned with the administration and growth of space to develop sustainable cities and communities. It can be differentiated into classes and types including integrated land use planning, participatory land use planning and regional land use planning (Metternicht 2018). Chigbu et al. (2017) and Filipe and Norfolk (2017) emphasise land use planning and output requirements such as the need for improving tenure security, or the need to address the poor. Ultimately, spatial planning can be viewed as a complex and iterative process with several internal feedback loops (Lagopoulos 2018; Wagner and de Vries 2019). In this study, spatial planning is viewed as an activity that creates a more coherent territorial organisation of land uses and linkages between them to balance demands with the need to protect environments and promote socio-economic development objectives (UN-Habitat 2008).

Spatial planning goals and processes from multiple integrated interdisciplinary perspectives are tied to the NUA. The achievement of socio-spatial equity and sustainable development is the central role of spatial planning (UN-Habitat 2009). It has a significant influence on the achievement of SDG 11 and the other 16 SDGs. It requires better spatial plans enabling affordable housing, transportation and energy resources for realisation of SDG 11 on sustainable cities and communities. It requires spatial plans that provide equitable framework for economic opportunities and access to jobs, public services, infrastructure, health and education facilities for realisation of SDG 1 on poverty reduction. Equally, it requires innovative spatial plans to provide a framework for clean water and sanitation under SDG 6.

4 Results

Spatial planning in Zimbabwe is directed by the Department of Spatial Planning and Development (DSPD), a technical arm of the government within the Ministry of Local Government and Public Works. The department has oversight over local authorities that must adhere to the provisions of the Regional, Town and Country Planning Act (RTCPA) (Chapter 29:12) and related legislation. In addition, the DSPD advises the Minister of Local Government and Public Works on the interpretation of planning provisions and the execution of planning functions under the reserved powers of the Minister, in accordance with the RTCPA. The department is also involved in facilitating sites for the development of government and parastatal activities and the detailed planning of estates on state land. "It also carries out technical evaluation of plans (master plans, local plans and layout plans) originating from local planning authorities to aid the Minister in making decisions on the plans, mediating in stakeholder conflicts arising from spatial planning decisions that interfere with other people's land rights" (Chirisa 2014; 6). Furthermore, the DSPD provides technical assistance relating to plan preparation structures for the development planning system under the Provincial Councils and Administration Act of 1985. It manages urban public transport, and carries out tacheometric survey and pegging of state land.

The achievement of socio-spatial equity and sustainable development is central to the role of spatial planning (UN-Habitat 2009). However, in Zimbabwe, the spatial planning system suffers from strong historical inertia and has not been effective in managing urban growth in ways that move forward sustainability and inclusion (Chipungu 2011; Rakodi 1995; Todes 2012; Toriro 2009; Watson 2016). During the colonial era, the spatial planning system was designed and operated for a tiny minority of the population (Gasasira et al. 1981). The colonial national and local spatial development model was discriminatory and served the needs of the minority. To this effect, the Regional, Town and Country Planning Act, 1996 was designed as a tool for managing urban development based on racial zoning. Efforts by post-colonial governments to de-racialise the city only managed to replace racially based zoning with a socially based (class) type of zoning and exclusion of poor communities (Munzwa and Wellington 2010). This colonial legacy is spelt out in more detail in Chap. 11.

The effectiveness of spatial planning in driving the NUA depends on enabling legislation and regulations. The centre piece spatial planning legislation, that is, the Regional, Town and Country Planning Act, 1996 is outdated and has lacked scope to address challenges arising from rapid urbanisation, gender, climate change and disasters (see Chap. 11, for an elaboration on these issues). Instead of providing for the development of sustainable and smart urban settlements that are safe, inclusive and resilient; the legislation retains a master planning and development control focus.

Master plans provide a framework for controlling development rather than steering it. Critics of the conventional master planning approach have called for redefinition of the spatial planning and the adoption of bottom-up strategic spatial planning which advocates for the adoption of pro-poor policies (Cities Alliance 2012; Patel et al. 2012). Strategic spatial planning does not address every part of a city but focuses only on those aspects or areas that are strategic or important to overall plan objectives. In this approach, planning is viewed as a strategic rather than comprehensive activity. This implies selectivity, and a focus on that which really makes a difference to the fortunes of an area or time. Master plans have further been criticised for reflecting little understanding of the dynamics of poverty and the peculiar nature of urbanisation in cities of developing countries (UN-Habitat 2009). By failing to accommodate informality, they contribute to social inequality.

In Zimbabwe, existing master plans are not responding adequately to the needs of the population in urban areas due to the fact that they are in pursuit of elitist colonial development codes (Chirisa et al. 2016). For example, the high standards of residential construction dictated by local authority by-laws reflect the past influence of established British building codes and practices, rather than a concern for the need to meet severe loading conditions. There has been little effort to modify these building regulations through the introduction of performance standards. In analysing the spatial planning system, no evidence was found of a spatial planning framework for effective implementation of the NUA.

4.1 National Urban Policy

Zimbabwe has prioritised implementation of the NUA through mainstreaming SDGs into national development plans, specifically the Transitional Stabilisation Plan (2018–2020) (Government of Zimbabwe 2018a) and the National Development Strategy 1 (NDS 1) (2021–2025). The national development plans have been complemented by the National Human Settlements Policy of 2020 and the Devolution and

Decentralisation Policy of 2020, which outline some strategies for the promotion of sustainable cities and communities, and the roles of tiers of government respectively.

The Transitional Stabilisation Plan had the objective of stabilising the macro-economic environment and set the country on a trajectory for economic recovery, but stabilisation remained elusive. The NDS 1 speaks more directly to the NUA by stipulating that as new settlements develop, focus will be on densification and development of integrated communities in order to minimise the cost of providing basic infrastructure and services. Further, the NDS1 sets targets to revamp infrastructure development which is critical to sustainable urbanisation. But the lack of a credible financing strategy for the NDS 1 has compromised the achievement of its objectives (Chirisa et al. 2020).

"Urban policy is a high-level coordinated guide that is prepared and implemented by government with other stakeholders to address urban issues with the aim to build productive, socially and environmentally sustainable cities" (Zimunya and Chirisa 2021; 1). It comprises vision, values and principles that represent the desired outcomes about urban land management, urban governance, service delivery and environmental sustainability (Glaeser 2012; Teitz 2015; USAID 2013). The Zimbabwe National Human Settlements Policy pronounces densification and urban regeneration as central strategies to be adopted in national urban policy. These strategies have been applied through residential infill development, reduction of stand sizes and provision of land for the construction of walk-up flats. In spite of efforts at densification, weak policy analysis has constrained socio-economic impact. A key informant felt that the lack of incentives for densification has stymied policy implementation and for cultural and economic reasons, residential flats have not been popular with people as they prefer a parcel of land of their own where they can engage in livelihood activities. The finding was confirmed by a representative of residents at Mbare infill flats, Harare who emphasised the need for urban design to integrate spaces for household use in food production.

National urban policies are particularly instrumental in achieving Target 11a, which states "Support positive economic, social and environmental links between urban, peri-urban and rural areas by strengthening national and regional development planning" (UN-Habitat 2019; 1). In Zimbabwe, there is no explicit policy framework to guide urbanisation. Yet urban policy provides a framework about the envisaged nature and quality of urban development and management as well as the governance structure that forms the basis of spatial planning (Cullingworth and Nadin 2006; Parker and Doak 2012; van der Berg 2019). Procedurally, planning strategies are prepared in line with the urban policy which precedes their formulation. In cases where urban policies and spatial plans are not interlinked, there may be negative implications on the quality of urban development.

4.2 Financial Devolution

Zimbabwe has been experiencing a slowdown in economic activity driven by the recession in 2019 and 2020, macro-economic instability, foreign currency and electricity shortages (Government of Zimbabwe 2020a). As a result, Zimbabwe's fiscal space has been inadequate to effectively support the implementation of the NUA. National government financial support to local authorities has been limited. Over the 1990–2019 period, central government allocations to sub-national governments rarely exceeded one percent of the total national budget making Zimbabwean local authorities mainly self-financed. In urban councils, the bulk of revenue goes towards funding recurrent expenditure. This is exacerbated by a debt-overhang by local authorities which constrains their capacity to access loans. Financial imbalances and capacity constraints have entailed that local authorities continue to experience challenges in maintaining efficient service delivery in key areas such as water and sanitation, road maintenance and refuse collection. There has been constrained financing for implementation of local authority spatial plans.

The 2013 National Constitution provides for devolution of power and authority to sub-national governments. In terms of the Constitution, at least 5% of the national revenue collections are allocated to lower tiers of government as inter-governmental fiscal transfers. In line with the Constitution, the Devolution and Decentralisation Policy was adopted in 2020 so as to among other things empower local authorities in spatial planning and development. But there seem to be inherent contradictions in the Constitution as it provides for devolution of decision-making powers and responsibilities to provincial and local authorities but maintains top-down resource allocation in the form of devolution funds (5% of the national revenue collections) (Box 1).

Box 1: Case of City of Masvingo and Devolution Funds

Another challenge noted during the research include interference by the central government in devolution funds. The permanent secretary in the Local Government and Public Works ministry sometime in July 2022 advised that government will procure fire tenders on behalf of local authorities utilising the 2022 devolution fund allocations. The City of Masvingo would get two fire tenders at a cost of US$464 396 each. The 2022 devolution allocation was $253 107 300 and council was being advised to re-prioritise the devolution funds to accommodate the cost of the fire tenders. The council officials highlighted that the fire tenders were not a priority to the City of Masvingo. "The City of Masvingo is in need of refuse trucks, tipper trucks, grader and ambulances, not fire tenders," said the official. The fire tender issue torched a storm after the Local Government ministry wrote to all city, town and rural district councils indicating that it was procuring fire tenders from Belarus on behalf of all local authorities with the money being deducted from their devolution fund allocations over a period of 12 months. "The directive by central government was unconstitutional since the funds are from devolution funds where they have to consult residents on how to use the money," said the mayor of Masvingo.

Source Hickonicko (2023: 8)

In its classical sense, devolution empowers sub-national governments to raise and retain a proportion of their revenue, and not the provision of devolution funds from the centre as in the Zimbabwe case. A senior manager in Bulawayo City Council amplified the matter when he mentioned that devolution funds are disbursed from central government in contradiction of the meaning of the concept. In his words "central government cannot come back and tell local the local council how to spend devolution funds. Devolution funds are disbursed in bits and pieces and if lucky the local council can get 30% of the allocation. Local councils are allocated devolution funds on condition that they implement certain types of projects and disbursement of funds has been slow" (Key informant). Hence the Constitution gives powers to sub-national governments with one hand and takes them away with the other. These policy contradictions have deprived local authorities a sound revenue base for funding spatial planning.

4.3 Corruption

Zimbabwe's Vision 2030 statement mentions that the country's aspirations for a prosperous and empowered upper middle-income society is anchored on an aggressive fight against all forms of corruption (Government of Zimbabwe 2018b). The Devolution and Decentralisation Policy document notes that efficient provision of services by local authorities has been compromised by corruption, particularly in the sale of land (Government of Zimbabwe 2020c).

The ability of spatial planning to steer the NUA is impaired by corruption in the land sector (Chirisa et al. 2020; City of Harare 2010; Matamanda et al. 2021). Land corruption is the abuse of power and authority by those in charge of land administration for their own gain or benefit (Chiweshe 2020). A special investigations report found that a senior politician manipulated procedures for the sale of state land in Harare for personal gain (City of Harare 2010).

Box 2: City of Harare: Case of Manipulation of Procedures for Personal Gain

Contrary to Harare City Council policy that an individual must not get more than one residential property from the Council, the Minister of Local Government, Public Works and National Housing (in charge of local authorities) acquired vast tracts of land within Greater Harare and registered them in companies associated with him. It remains disturbing to note that the Minister would identify pieces of land in the city, influence Council Officials to apply to him for change of land use, and then preside over the same applications and approve the changes. He would then write to Council Officials asking to buy the same stands and obviously get them. Land reserved for recreational activities would end up having Title Deeds in his company's name. A case in point is Stand 61 Hellensvale Harare, measuring 20 hectares. According to the Advice of Payment, the Minister paid $2300 for this stand.

Source City of Harare (2010: 8–9).

Councils are guided by the Urban Councils Act Chapter 29:15, and its regulations. Section 152 (2) of the Urban Councils Act requires that before alienating Council land, that is, selling, exchanging, leasing, donating or otherwise disposing of or permitting the use of any land owned by it; Council shall, by notice published in two issues of a newspaper advertisement, inform the public of its intention to do so. The notice gives the public the right to object to the Council's proposal. The report found that in all cases investigated, the requirements of Section 152 of the Urban Councils Act were not followed resulting in anomalies and irregularities in the alienation of Council land. Policies and procedures on Council land alienation were at times completely ignored. The report makes reference to rampant corruption in all departments of council, with the department of urban planning services worst affected by the scourge. Nationwide, corruption has been frustrating the NUA with executives and councillors in local authorities failing to address urban development challenges (Chirisa et al. 2020; Matamanda 2021).

Corruption in spatial planning goes beyond planners to encompass politicians, councillors and citizens. In urban areas partisanship has often resulted in large land parcels being transferred to individual land developers rather than to urban councils thereby undermining potential revenue which could be used by local authorities to provide infrastructure and services (Key informant). This scenario has contributed to the marginalisation of spatial planning in the political economy of urban development. Corruption within the socio-political leadership, including the complication of political factions, has made it difficult to lead towards a sustainable urban agenda.

Box 3: Reflections by a Former Director of Physical Planning

At the professional and technical levels, we have lost it through corruption. There are negative trends within the planning profession. For example, there has been a case where one finds stands being created in a road reserve. People bring development proposals together with a khaki envelope (cash bribe). There are elements of corruption within the socio-political leadership and complication of political factions which makes it difficult for leaders to lead on rational and long-term benefits. There have been residential developments that were informally authorised.

Note The interview was conducted on September 9, 2022, and the interviewee requested that the quote be anonymous.

A key informant reckoned that many councillors are not there to serve but to benefit from the urban council system. The corruption scourge is pursued in more detail in Chap. 11.

4.4 Informal Settlements

Since 2000, political and economic crisis led to a sharp increase in urban informality and this has often directly contributed to social and spatial marginalisation. The effects of economic decline have been exacerbated by the global recession which began in 2008 and has brought about economic structuring and rapid growth of unemployment in all parts of the world. In the constrained financing context, the limited performance of municipal budgets has resulted in huge infrastructure and service backlogs (Government of Zimbabwe 2020b). Meanwhile, the urbanisation of poverty has reduced the capacity of communities to contribute revenue towards service provision (Chanza and Chirisa 2011; Machakaire 2015). The problem clearly manifests in peri-urban farms that were acquired by central government for urban development and have mostly been developed by politically aligned cooperatives with minimal provision of infrastructure and services. The proliferation of informal settlements was catalysed by central government's parallel development policy of 2006 that allowed property developers to raise funds for provision of services by selling off stands before infrastructure had been put in place. But the realised funds were rarely used for their intended purpose. The informal settlements which have emerged in peri-urban areas are generally characterised by a chronic lack of basic infrastructure services, inferior housing, illegal dwellings, illegal or insecure land tenure, high construction density, poor sanitary conditions, poverty, social exclusion, self-production, and incremental development (Dovey et al. 2020; Huchzermeyer 2003).

Previous central government policies emphasised demolition of informal settlements but the NDS 1 and the Zimbabwe Human Settlements Policy promote their upgrading (Muchadenyika 2015; Tibaijuka 2005). The upgrading of informal settlements would make them safer, liveable, resilient and sustainable. But the sustainability of informal settlement upgrading in view of the rapid urbanisation challenge in Zimbabwe and other developing countries has been questioned. Informal settlements do not yield much revenue for local authorities yet their upgrading has been empirically found to cost three times more than site and services schemes (Chirisa 2010). Nonetheless, nationwide pilot projects involving partnership between the Zimbabwe

Homeless People's Federation, non-governmental organisations, communities and urban councils have shown promise to deliver sustainable informal settlements provided the government funds trunk infrastructure. As well, in line with national policies, there has been a shift from demolition to upgrading of informal settlements at the level of urban councils. For instance, Harare City Council has developed an informal sector development policy and citywide slum upgrading programme. This initiative has strengthened access to economic opportunities for women, the poor and other vulnerable groups. Despite the absence of a legal and institutional framework, the adoption of informal settlements upgrading is in conformity with the tenets of the NUA (Mehatty and Haas 2020).

4.5 Regulation of Planners

Pursuant to the corruption scourge that has been discussed above, one of the concerns that have been raised by the planning fraternity in Zimbabwe is that the spatial planning profession is not regulated (Masiya and Mwedzi 2018; Matamanda et al. 2021). In the absence of a Planners Act, practicing planners are not registered and the Zimbabwe Institute of Regional and Urban Planners (ZIRUP) is a charitable organisation without a legal basis for enforcing the code of ethics. In the absence of a statutory framework for regulating planning professionals in the country, the Zimbabwe Institute of Regional and Urban Planners has not regulated the education and training of planners—thereby missing the opportunity to strengthen the role of spatial planning in steering the NUA. Further, ZIRUP has no power to reprimand or sanction its members and a planner can withdraw membership with no consequences for continued practice (Mukarwi et al. 2018).

When planning is a profession, planners belong to an Institute which would be a key stakeholder in ensuring responsiveness of spatial planning legal frameworks to the needs of citizens. The registration of planners would be a requirement for them to practice in the profession and they would be subject to a code of conduct. The regulatory body sets standards for entry, practice, and removal of individuals from the register. Members found guilty of violating the code of conduct can be reprimanded, suspended or get their practicing licence revoked. Hence the registration of planners would protect both the profession and the public. In the absence of a requirement for registration of planners, professionalism has suffered because there have been no effective checks against corrupt practices (Kamuzhanje 2018). It is accepted that regulation of planners by itself may not fully address the challenges facing the planning profession, but it nonetheless contributes to an improvement in the planning system.

5 The Emerging Built Environment Focus

In spite of the aforementioned spatial planning challenges, the discipline anchors the built environment focus. It creates a base or framework on which other built environment disciplines like land surveying, civil engineering, quantity surveying, architecture and real estate management add value. Traditionally, built environment sciences at the University of Zimbabwe have operated as closed guilds; but a change in curriculum to a built environment inter-discipline, which offers some common courses, would strengthen the disciplines through multi-skilled products for driving the NUA. This would constitute a step towards a built environment concept. Such an approach would require all built environment students to study some common courses at least in the first year or give them opportunity to enrol for a limited number of courses in other built environment disciplines.

The spatial planning curriculum at the University of Zimbabwe is robust in that apart from traditional planning courses such as regional planning, urban planning and urban design, students go beyond and also study economics, accountancy, construction economics, building technology and property valuation. This has not only produced a multi-skilled planner appropriate for a developing country context but has opened opportunities for planners to venture into real estate management, valuation and quantity surveying.

6 Conclusion and Recommendations

The chapter found that spatial planning provides the framework that enables other disciplines and practices to add value to the built environment. It has been contributing to the NUA through translating the objectives of national policies such as the National Development Strategy 1 (2021–2025) and the Zimbabwe Human Settlements Policy (2020) into space. Notable progress in spatial planning has been recorded in the change in policy from demolition to upgrading of informal settlements, a shift that has enabled more sustainable, resilient and inclusive cities. However, a number of challenges such as the lack of a national urban policy, incoherent financial devolution, land corruption and non-registration of planners have undermined the potential of spatial planning to steer the NUA. The way forward should build on efforts that have already been shown towards more inclusive cities while addressing shortcomings that have been identified in the spatial planning system.

The following are recommendations on how spatial planning in Zimbabwe can be shaped to more effectively steer the NUA (Table 1).

It is noted that revamping the spatial planning system in Zimbabwe is a substantial task given the political and economic crisis, but attention to the recommendations would address some key areas such as those relating to national urban policy and corruption. Needless to mention that spatial planning works in a context of political

Table 1 Action plan

Spatial planning challenge	Needed action	Responsible agent
1. Spatial planning system has a development control focus	Adoption of strategic spatial planning and pro-poor policies	Ministry of local government and public works
2. No national urbanisation strategy for steering the NUA	Develop a national urban policy	Ministry of local government and public works
3. Arbitrary disbursement of devolution funds to local authorities	Develop a devolution law that streamlines structures and procedures for devolution	Ministry of finance and economic development
4. Land corruption	Establish safeguards against improper use of spatial planning powers	National leaders
5. Unregulated planners	Develop a planners act	Ministry of local government and public works

and macro-economic stability, effective government, regulatory quality, rule of law and control of corruption (Kaufmann et al. 2007).

References

Caprotti F, Cowley R, Datta A, Broto VC, Gao E, Georgeson L, Herrick C, Odendaal N, Joss S (2017) The new urban agenda: key opportunities and challenges for policy and practice. Urban Res Pract 10(3):367–378

Chanza N, Chirisa I (2011) People or environmental quality? Solid waste management dilemma in Epworth, Harare. United Nations Centre Reg Dev Reg Dev Stud 15:42–57

Chigbu UE, Schopf A, De Vries WT, Masum F, Mabikke S, Antonio D, Espinoza J (2017) Combining land-use planning and tenure security: a tenure responsible land-use planning approach for developing countries. J Environ Plann Manage 60(9):1622–1639

Chipungu L (2011) Insights into urban development control challenges: a case study of operation Murambatsvina/restore order in Zimbabwe. Built Human Environ Rev 4:17–32

Chirisa I (2010) Peri-urban dynamics and regional planning in Africa: implications for building healthy cities. J Afr Stud Dev 2(2):15–26

Chirisa I (2014) Building and urban planning in Zimbabwe with special reference to Harare: putting needs, costs and sustainability in focus. Consilience: J Sustain Dev 11(1):1–26

Chirisa I, Maphosa A, Zanamwe L, Bandauko E, Mukarwi L (2016) Past, present and future population growth and urban management in Zimbabwe: putting institutions in perspective. In: Benna UG, Bello AS, Garba SB (eds) Population growth and rapid urbanisation in the developing world. IGI Global, Hershey, Pennsylvania

Chirisa I, Matamanda AR, Mazanhi P (2020) Resisting, frustrating or embracing the urban agenda: Chieftaincies in Southern Africa examined constitutionally and statutorily. Land Use Policy 95:1–7

Chiweshe MK (2020) Money, power and the complexities of urban land corruption in Zimbabwe. Afr J Land Policy Geospatial Sci 3(5):131–141

Cities Alliance (2012) Guiding urbanisation in Africa for economic development and poverty education. ISTED
City of Harare (2010) Special investigations committee's report on city of Harare's land sales, leases and exchanges from the period October 2004 to December 2009. Councillor W. Dumba
Cullingworth B, Nadin V (2006) Town and country planning in the UK. Routledge, New York
Dovey K, Van Oostrum M, Chatterjee I, Shafique T (2020) Towards a morphogenesis of informal settlements. Habitat Int 104:102240
Filipe E, Norfolk S (2017) Changing landscapes in Mozambique: why pro-poor land policy matters. IIED Briefing Paper-International Institute for Environment and Development, London
Glaeser EI (2012) The challenge of urban policy. J Policy Anal Manage 31(1):111–122
Gasasira F, Amos F, Carol J, McAuslan P, Daniel J (1981) A spatial planning system for Zimbabwe, Harare, UNDP/UNCHS
Government of Zimbabwe (2018a) Transitional stabilisation programme reforms agenda October 2018-December 2020 'towards a prosperous and empowered upper middle-income society by 2030'. Harare, Government Printers
Government of Zimbabwe (2018b) Zimbabwe vision 2030 'towards a prosperous and empowered upper middle-income society by 2030'. Harare, Government Printers
Government of Zimbabwe (2020a) Towards a prosperous and empowered upper middle-income society by 2030,' national development strategy 1, January 2021-December 2025, Harare, Government Printers
Government of Zimbabwe (2020b) Zimbabwe national human settlements policy, ministry of local government, public works and national housing, Harare
Government of Zimbabwe (2020c) Devolution and decentralisation policy, Harare, Government Printers
Hickonicko B (2023) Devolution and public service delivery: evidence from Masvingo City, Zimbabwe. J Urban Syst Innov Resilience Zimbabwe 5(1):1–16
Huchzermeyer M (2003) Housing rights in South Africa: invasions, evictions, the media, and the courts in the cases of Grootboom, Alexandra, and Bredell. Urban Forum 14(1):80–107
Kamuzhanje J (2018) The capitulation of professional planning practice- myth, mystery or reality: an external observer's balance score card on professional planning practice in Zimbabwe. In: A paper presented at the Zimbabwe Institute of Regional and Urban Planners Annual School, 28–30 August 2018, Troutbeck Resort, Nyanga
Kanyenze G (2021) Leaving so many behind: the link between politics and the economy in Zimbabwe. Weaver Press, Harare
Kaufmann D, Kraay A, Mastruzzi M (2007) Worldwide governance indicators project. World Bank Policy Research Working Paper 4149, Washington, DC
Lagopoulos A (2018) Clarifying theoretical and applied land-use planning concepts. Urban Sci 2(1):1–20
Machakaire DG (2015) Transformation of urban planning practices using geo-spatial technology in managing rapid urbanisation in Harare: Zimbabwe (Master of Technology dissertation, Cape Peninsula University of Technology)
Masiya JM, Mwedzi TP (2018) The institutional set up and challenges of planning in Zimbabwe. In: A paper presented at the Zimbabwe Institute of Regional and Urban Planners Annual School, 28–30 August 2018, Troutbeck Resort, Nyanga
Matamanda AR (2021) Mugabe's urban legacy: a postcolonial perspective on urban development in Harare, Zimbabwe. J Asian Afr Stud 56(4):804–817
Matamanda AR, Chirisa I, Leboto L, Dunn M, Kwangwama A (2021) The planning profession questioned: evidence from the role and practice of planners in Zimbabwe. J Plan Educ Res 1–7
Mehaffy M, Haas T (2020) New urbanism in the new urban agenda: threads of an unfinished reformation. Urban Plann 5(4):441–452
Metternicht G (2018) Land use and spatial planning: enabling sustainable management of land resources. Springer

Muchadenyika D (2015) Slum upgrading and inclusive municipal governance in Harare, Zimbabwe: new perspectives for the urban poor. Habitat Int 48:1–10
Mukarwi L, Chirisa I, Maphosa A (2018) Real estate and planning professional practice. In: Chirisa I (ed) Fundamentals of planning and real estate studies: a primer for Zimbabwe. UZ Publications, Harare, pp 435–452
Munzwa K, Wellington J (2010) Urban development in Zimbabwe: a human settlement perspective. Theor Empir Res Urban Manag 5(14):120–146
Oppong C (2006) Familial roles and social transformations: older men and women in Sub-Saharan Africa. Res Aging 28(6):654–668
Parker G, Doak J (2012) Key concepts in planning. SAGE Publications, London
Patel S, Baptist C, D'Cruz C (2012) Knowledge is power-informal communities assert their right to the city through SDI and community-led enumerations. Environ Urban 24(1):13–26
Rakodi C (1995) Harare: inheriting a colonial city, change or continuity? John Wiley, Chichester and New York
Stead D, Nadin V (2008) Spatial planning: key instrument for development and effective governance with special reference to countries in transition. United Nations, Geneva
Teitz MB (2015) Urban policy in North America. International encyclopaedia of the social and behavioural sciences. USA, Elsevier, Cambridge, pp 908–914
Tibaijuka AK (2005) Report of the fact-finding mission to Zimbabwe to assess the scope and impact of operation Murambatsvina. United Nations, Geneva
Todes A (2012) New directions in spatial planning? Linking strategic spatial planning and infrastructure development. J Plan Educ Res 32(4):400–414
Toriro P (2009) The impact of the economic meltdown on urban agriculture in Harare. Urban Agric Mag 21:26–27
UN-Habitat (2008) State of the world's cities 2008/2009: harmonious cities. United Nations Human Settlements Programme, Nairobi, Kenya
UN-Habitat (2009) Planning sustainable cities: global report on human settlements. Earthscan, London
UN-Habitat (2016) Urbanisation and development: emerging futures, world cities report 2016, Nairobi. United Nations Human Settlements Programme
UN-Habitat (2019) How to formulate a national urban policy: a practical guide. United Nations Human Settlements Programme, Nairobi, Kenya
USAID (2013) Sustainable service delivery in an increasingly urbanizing world, USAID policy, USAID
Van der Berg A (2019) Can South African planning law and policy promote urban sustainability in the Anthropocene? In: M Lim (ed) Charting environmental law futures in the Anthropocene, Springer Nature, Singapore, pp 203–220
Wagner M, de Vries WT (2019) Comparative review of methods supporting decision-making in urban development and land management. Land 8(8):1–3
Watson V (2016) Locating planning in the new urban agenda of the urban sustainable development goal. Plan Theory 15(4):435–448
Wekwete K (2018) Planning and real estate fundamentals: an introduction. In: Chirisa I (ed) Fundamentals of planning and real estate studies: a primer for Zimbabwe. University of Zimbabwe Publications, Harare, pp 1–9
Zimunya W, Chirisa I (2021) Urban policy and the future of urban and regional planning in Africa. In: Brears R (ed) The Palgrave encyclopaedia of urban and regional futures. Palgrave Macmillan, pp 1–11

Construction and Civil Engineering Steering the New Urban Agenda in Zimbabwe

Samson Shumba, Pride Mutekwa, and Prosper Marindiko

Abstract The study focuses on the contribution of construction and civil engineering to the New Urban Agenda (NUA) in Zimbabwe and its contribution to the sustainable development goals (SDGs). The NUA is endorsed by the UN member states. A review of the progress in implementing the NUA in Zimbabwe and the contribution of the government, academia and industry was conducted. The study also covered the assessment of the contribution of professional bodies for the attainment of the NUA goals and the contribution of Education 5.0 through teaching, community outreach, research, innovation and industrialisation. Geotechnical, water quality and materials investigations were conducted as well as questionnaires and key informant interviews for providing the indicators for performance. Analysis of results was done using quantitative assessment tools as well as appropriate engineering software. Case studies on water and sanitation, integrated agro-processing plants, markets and smart settlements and the environment were done. The study concludes that, there is significant progress in addressing the main issues as evidenced by the models and infrastructure developed. The research recommends an implementation plan for the success of the NUA in Zimbabwe and Sub-Saharan Africa.

Keywords New urban agenda · Infrastructure · Industrialisation · Construction and civil engineering · Built environment · Sustainable development · Water · Sanitation and hygiene

S. Shumba (✉) · P. Mutekwa · P. Marindiko
Department of Construction and Civil Engineering, University of Zimbabwe, Harare, Zimbabwe
e-mail: sshumba@eng.uz.ac.zw

C. Chavunduka and I. Chirisa (eds.), *New Urban Agenda in Zimbabwe*,
Advances in 21st Century Human Settlements,
https://doi.org/10.1007/978-981-97-3199-2_6

1 Introduction

1.1 Background

The New Urban Agenda (NUA) recognises the role of cities in achieving the SDGs (Valencia et al. 2019). World leaders met with urban development partners and scientists at the Habitat III conference in Quito to adopt the NUA (Garschagen et al. 2018). The NUA is voluntary in implementation despite its endorsement by the UN General Assembly. Some scholars argue that achievement of the NUA goals is unrealistic (Garschagen et al. 2018). Social, economic and environmental issues have to be addressed by the New Urban Agenda and the sustainable development goals in Zimbabwe. The NUA was formulated by the United Nations as a way of achieving the sustainable development goals by governments (implementing these through local authorities).

In the light of this background, the study seeks to focus on the contribution of construction and civil engineering to the achievement of the NUA in Zimbabwe as well as sustainability and better way of living in the nation. Focus is made on infrastructure provision, Water, sanitation and hygiene (WASH) and industry. The study will investigate the adoption, implementation and review of the New Urban Agenda in Zimbabwe construction industry.

1.2 Overview

In Zimbabwe, the sustainable development goals are implemented through the Ministry of Public Service, Labour and Social Welfare. Vision 2030 (Towards a Prosperous and Empowered Upper Middle-Income Society by 2030) outlines the development agenda for Zimbabwe thereby addressing some of the key challenges in implementing the NUA. Vision 2030 addresses the following key challenges in construction and civil engineering: irrigation development, transportation (urban mass transit system, dualisation and ring roads, rail ports of entry), airport infrastructure, water and sanitation and housing.

The SDGs and the NUA are ambitious, comprehensive socially progressive agendas which have the potential to contribute to the transition towards more sustainable, inclusive and resilient cities by serving as tools that question the status quo and mobilise actors and resources. The NUA offers solutions to global urban problems. Some of its targets include "helping to end poverty and hunger in all its forms and dimensions", "reducing inequalities", "promoting sustained, inclusive and sustainable economic growth", "achieving gender equality and the empowerment of all women and girls in order to fully harness their vital contribution to sustainable development", "improving human health and well-being, fostering resilience and protecting the environment". The major obstacles in achieving the NUA are lack of trust between public and private sectors, lack of partnering skills and capabilities and absence of support to facilitate complex collaboration (Garschagen et al. 2018). The NUA is critical for nations in achieving a better and more sustainable future.

1.3 Review of Progress

The New Urban Agenda outlines a review process towards the implementation of the SDGs (Garschagen et al. 2018). This includes the development of local follow-up and review mechanisms by local governments and their partners and the role of the World Urban Forum in the follow-up and review process. The quadrennial implementation report will feed into the General Assembly and High-level Political Forum to support coherence, coordination and collaborative linkages in the follow-up and review of the 2030 Agenda for Sustainable Development and NUA.

2 Literature Review and Theoretical Framework

2.1 Industry Contribution to the New Urban Agenda

Construction industry is an umbrella term covering companies involved in the building of structures (Ebohon and Rwelamila 2001). Industries involved in the construction sector have a very large contribution to make to the NUA as they are the drivers and benefactors of NUA (Ofori et al. 1996). Industries involved in building construction are diverse from design, construction, project management and maintenance. They all have major contributions to the NUA albeit at different stages or in different ways which ultimately lead to a sustainable preservation of the Zimbabwe's resources for future generations. The NUA is contained in sustainable development goal number 11 which mainly deals with sustainability in cities, and thus, the construction industry has to align to this or be left behind (Watson 2016).

The Zimbabwean construction industry greatly contributes to the NUA by making all its activities sustainable and deliberately taking measures to protect and conserve the environment through reducing the appetite of their activities for more raw materials. Those involved with the design aspect of construction can deliberately design less material and energy demanding structures which are known as green buildings. Green buildings are designed and constructed to incorporate features like low flush toilets, motion sensitive lighting, use of energy rated products, use of renewable materials, use of paints which do not emit volatile organic compounds (VOC's), greywater reuse, use of brownfields, preservation of the natural environment (Zuo and Zhao 2014) among several others which are in sync with the New Urban Agenda.

While the construction industry has a direct role to play in the implementation of sustainable construction policies there are other players who also have a role to play in the New Urban Agenda. Policy makers and engineering bodies like the Zimbabwe

Institution of Engineers (ZIE) and the Engineering Council of Zimbabwe (ECZ) have a role to play in pushing the NUA. Academic institutions like universities and the Scientific and Industrial Research and Development Centre (SIRDC) also have a role to play through their innovation hubs and research acumen in promoting sustainable Zimbabwean construction standards. Achievement of the NUA is a whole life cycle operation which can only be via the participation of all players as a neglect of one aspect during the design stage for example might lead to unsustainable operational costs or waste production during the building life.

The construction and civil engineering industry is governed by numerous professional bodies which are responsible for ensuring that construction professionals maintain adequate standards of training and ethics. They also support and promote professional learning for students and practitioners safeguarding the status of the professionals and representing their interests in dealings with the government and other bodies. These professional bodies include Zimbabwe Institution of Engineers (ZIE), Engineering Council of Zimbabwe (ECZ), and the Construction Industry Federation of Zimbabwe (CIFOZ).

2.2 New Urban Agenda and the National Development Goals

Central and local governments with their institutions are responsible for the delivery of basic infrastructure such as water, sanitation, electricity, solid waste collection, and social services (education, health and welfare) and the built environment (human settlements). They have the mandate for infrastructure provision and maintenance. For the city to successfully meet its goals and achieve sustainability the national and international goals should be related. Some of the major challenges that the city has limited capacity to address include issues around social development, education, health, as well as safety and security, all of which have been receiving dwindling central government funding. Awareness and engagement with regard to the SDGs is growing in cities. Developing interlinkages between global agendas is crucial. There is need to consider the range of interdependencies among the global agendas as significant to unlocking their full potential, optimising resources and leveraging resource development outcomes at every level of government.

The Government of Zimbabwe is making significant strides towards the implementation of the New Urban Agenda. Numerous activities are underway as the government aims towards a prosperous and empowered upper middle-income society by 2030, which encompasses all SDGs. To date the government has launched a number of programmes which will accelerate the realisation of the NUA. These include:

i. ***National Development Strategy 1 and Zimbabwe Vision 2030***

As the National Development Strategy 1 (NDS1) is implemented, Government will move with the sustainable development goals (SDGs) theme of "Leaving no-one

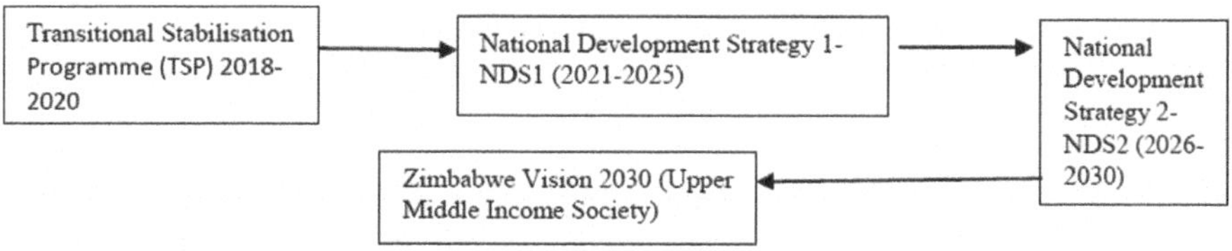

Fig. 1 Three phase process to attain the country's Vision 2030 Outcomes. *Source* Government of Zimbabwe (2020)

and no place behind" (Government of Zimbabwe 2020). One of the macroeconomic objectives for the 5-year period (2021–2025) of the NDS1 is as follows:

- To improve infrastructure development and investment in energy, water, sanitation, roads, health, housing, education and social amenities.

During the NDS1, efficient infrastructure delivery will be key in the realisation of the national priorities and overall socio-economic development. This will involve restoring basic infrastructure in transport, water, sanitation and housing (Government of Zimbabwe 2020). Figure 1 shows the development trajectory towards the realisation of the Upper Middle Income society by 2030.

NDS1 is the first 5-year medium-term plan aimed at realising the country's Vision 2030. The national priorities identified are housing delivery, infrastructure and utilities. Table 1 shows the relationship between VISION 2030, SDGs and National Priorities (Government of Zimbabwe 2020).

The delivery of affordable, sustainable, modern, functional housing and social amenities in urban and rural areas will be the government's priority under NDS1. Rental accommodation such as flats will be increased. NDS1 targets a cumulative 220,000 housing units to be delivered by 2025 in rural and urban areas. The number of kilometres of road network in good condition will be increased from 14,702–24,500 km by 2025. The Government of Zimbabwe has managed to deliver on some of the priority projects that provide downstream benefits to the economy and these include but are not limited to:

- Tugwi Mukosi dam,
- Bulawayo and Victoria Falls airport upgrade,
- Beitbridge Border Post modernisation, and
- Road rehabilitation projects.

Table 1 Relationship between VISION 2030, SDGs and national priorities

VISION 2030 pillar	SDG	National priority
Inclusive growth	SDG 9, 10 and 11	Infrastructure and utilities
Infrastructure and utilities	SDG 9 and 11	Housing delivery

Source Government of Zimbabwe (2020)

ii. ***Emergency Road Rehabilitation Programme (ERRP)***

The Emergency Road Rehabilitation Programme (ERRP) is a programme by the Government of Zimbabwe to repair and construct the road network in the country. The ERRP will be implemented in three phases, that is, ERRP I, ERRP II and ERRP III. The programme has been phased to cover rehabilitation of major highways, resurfacing, pothole sealing, shoulder widening and drainage renovations and the final stage will target feeder roads and roads in communities. Currently ERRP Phases 1 and 2 are underway. There are projects in Harare, KweKwe, Gweru and other urban centres including rural district councils.

2.3 Challenges to Achieving the NUA

2.3.1 New Urban Agenda and Infrastructure

The starting point for achieving the NUA and sustainability is infrastructure development. Lack of infrastructure holds back development. Planning systems need to encompass the powers necessary to co-ordinate the investment of essential infrastructure. Implementing the New Urban Agenda in order to achieve the sustainable development goals always includes the provision of sustainable infrastructure. There is a need to invest in climate sensitive, inclusive and equitable solutions, be it carbon–neutral transportation, efficient waste management and recycling systems, heating, housing or almost any other infrastructure project to be launched.

By 2050, the world's urban population is expected to nearly double, making urbanisation one of the twenty-first century's most transformative trends. Populations, economic activities, social and cultural interactions, as well as environmental and humanitarian impacts, are increasingly concentrated in cities, and this poses massive sustainability challenges in terms of infrastructure. Infrastructure provision is among the greatest drivers of cost and resource efficiencies, through the benefit of economies of scale and agglomeration and by fostering energy efficiency, renewable energy, resilience, productivity, environmental protection and sustainable growth in the urban economy.

Listed below are some of the major infrastructure projects that the Government of Zimbabwe is actively implementing under the NUA.

i. Tugwi-Mukosi Dam Project,
ii. Gwayi-Shangani Dam Project,
iii. New City Mount Hampden Project,
iv. Harare-Beitbridge Road,
v. Tertiary Institutions Infrastructure Development Projects, and
vi. National Housing Delivery Project.

2.3.2 New Urban Agenda and Water, Sanitation and Hygiene

The Water, Sanitation and Hygiene (WASH) policy promises universal and equitable access to safe and affordable drinking water for all; as well as access to adequate and equitable sanitation and hygiene for all; and an end to open defecation, with special attention to the needs and safety of women and girls and those in vulnerable situations.

New Urban Agenda calls for equipping public water and sanitation utilities with the capacity to implement sustainable water management systems, sustainable maintenance of urban infrastructure services, (...and promote) universal and equitable access to safe and affordable drinking water, and adequate and equitable sanitation and hygiene for all.

During the 1980s and 1990s period, the Urban WASH coverage in Zimbabwe was almost 100% with a very high level of service (UNICEF Zimbabwe 2018). However, a sharp decline was experienced in the late 1990s due to the political and economic crisis which left the WASH service providers with no foreign exchange to procure spare parts and water treatment chemicals. Moreover, there was very minimal budget allocation on water and sanitation from the central government which crippled Local Authorities. Thus, they could not undertake their mandate of building, operating, upgrading and maintaining water infrastructure including the conveyancing and reticulation systems (Government of Zimbabwe 2020). Additionally, unreliable energy supplies also affected efficient operation of the water supply and sewerage systems in urban areas. These events consequently led to a general distrust between residents and the local authorities resulting in a low willingness to pay, thus failure to generate revenue for service provision. This was followed by an exodus of trained WASH experts in search of better opportunities, severely affecting the operations and maintenance of WASH systems. Furthermore, under investment for expansion and upgrade of the systems inevitably resulted in the insufficiency of the infrastructure to cater for the needs of an ever-increasing urban population with only 29.7% of households having access to improved water sources and sanitation (Government of Zimbabwe 2018). The compromised water supply system with intermittent supplies forced the residents to look for alternative water sources. This resulted in low sewage flows subsequently leading to blocked sewers. In some high-density suburbs, raw sewage ended up flowing in streets.

3 Methodology

A mixed methods approach was used to collect data through interviews and questionnaires. The study adopted both primary and secondary data sources. A questionnaire was administered on the adoption, implementation and review of the NUA in Zimbabwe with the aim of assessing the extent of contribution of construction and civil engineering organisations and professionals. Case studies were used to show the role of engineering in the built environment and how it has helped to solve problems being faced by the community. The larger part of the research was done through a comprehensive literature review.

Responses were obtained from professionals aligned to the Government (25%), Academia (10%), Private (55%) and Non-Governmental Organisations (10%). The respondents were from a pool of expertise in Energy, Water and Sanitation, Sewer Reticulation, Roads, Irrigation, Structural Engineering, Construction, Engineering consultancy and Project Management. Statistics showed that 35% of the respondents were knowledgeable about the New Urban Agenda whereas the remaining 65% were clueless on the issue.

4 Results

Based on the gathered data, the NUA is a viable programme which creates a participatory approach which brings acceptability, sense of ownership, and if well implemented greatly improves the social and economic livelihoods of urban and rural communities in a transparent way. The goals are transformative through social inclusion in development, however there is need for more awareness programmes in Zimbabwe. The activities in the urban spaces of the world, especially in developing countries, require a lot of resources and energy due to skyrocketing population. To date, NUA goals have been overwhelmed by the slow pace of development hence the need for acceleration and reconstruction. A robust implementation of the NUA in Zimbabwe will make urban settlements safer and bring sustainability to the infrastructure.

Although there were concerns by some that the goals of the NUA are unrealistic and unachievable, most respondents showed interest in the NUA and indicated that if the Government of Zimbabwe was to give priority to the initiative, sustainable urban settlements can be realised. About 50% of the respondent organisations are involved in work related to the New Urban Agenda and have driven WASH related projects in areas such as Chipinge and Chimanimani, rehabilitated and constructed new road networks in different cities and towns nationwide and some have been actively involved in the construction of state-of-the-art structures including shopping malls. Moreover, professionals are also contributing in a significant way towards the realisation of the NUA. New building technologies that are sustainable and resilient have been introduced. Some are directly involved in urban planning and development whilst others are sitting in decision making boards and endorsing feasibility of government's infrastructure projects.

In the sections that follow, we provide cases of work by central government, local government, development partners, non-governmental organisations, the private sector and academia towards implementation of the NUA. The cases span industrial, road, water, sanitation and hygiene infrastructure and highlight central government commitment to use the NDS1 as one of the vehicles for realising targets set under the NUA.

4.1 New Urban Agenda and Industry and Road Infrastructure

4.1.1 Design of Fruit and Vegetable Processing Plant

Zimbabwe is an agro-based economy and agriculture is the mainstay of the country. The government launched the command agriculture initiative to increase food security and economically empower rural communities (Mapangire et al. 2019). The project sought to address the need for horticultural produce processing factories around the farms, thus, an integrated vegetable and fruit-processing factory structure was designed for Juru Service Centre in Goromonzi District, Mashonaland East Province, Zimbabwe. In-depth quantitative and qualitative research provided key data on the type of vegetables and fruits grown by farmers in Goromonzi District (Mapangire et al. 2019).

The geotechnical site and laboratory investigations carried out indicated that, the in situ soil had satisfactory properties for supporting the structure without any improvements. For structural analysis, finite element analysis software ANSYS was used. The factory layout design was based on an operational flow through pattern principle with sufficient space to allow concurrent processing of different types of horticulture products. Methane gas and bio-fertiliser are some of the by-products which results in 30% cost saving on the farmers (Mapangire et al. 2019).

4.1.2 Redesign of Simon Mazorodze and Chitungwiza Road Intersection in Harare

This project focussed on analysing and redesigning of Simon Mazorodze/ Chitungwiza road intersection in Harare, Zimbabwe. Different types of intersections and the cost-effective ways of improving roundabout intersections were analysed using standard procedures. A 96-h manual traffic survey was conducted at the intersection to determine current traffic volumes, intersection capacity and performance (Philip et al. 2017). The recommended design was of a roundabout with bypass slip lanes (shown in Fig. 2) and a design life of 20 years (Philip et al. 2017). The design involved introducing adjacent bypass slip lanes to all approaches of the existing roundabout and widening of entry and exit width. An overpass was recommended to cater for pedestrians.

4.1.3 Construction of Low Volume Roads Using Labour-Based Technology: Case of Mutare

There is a huge backlog in the construction and upgrading of roads in Zimbabwe. This is much pronounced in urban residential areas due to the lack of readily available and relatively cheap road-building materials (Mukura et al. 2018). The project sought

Fig. 2 Design solution of a roundabout with slip lanes. *Source* Philip et al. (2017)

to design the pavement structure for low-volume roads, Dangamvura Extension, Mutare. Extensive geotechnical investigations, traffic counts and forecasting, as well as studies of suitable design parameters for pavement structure and requirements in applications of labour-based technology were carried out for valuable conclusions and beneficial recommendations to residential estate development (Mukura et al. 2018). The innovation lies in minimising the construction cost of roads in urban residential areas without compromising the standards for road construction, through the use of sand which requires less processing as a sand seal, combined with the use of labour-based technology. Figure 3 shows a 15 m road reserve cross section of the pavement designed using sand seal.

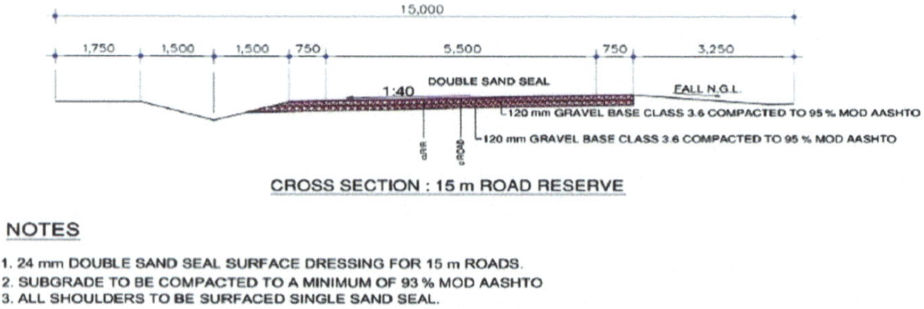

Fig. 3 A 15 m road reserve cross section using sand seal. *Source* Mukura et al. (2018)

4.2 *New Urban Agenda and Water, Sanitation and Hygiene*

4.2.1 Integrated Urban Water Management

The NUA includes Integrated Urban Water Management (IUWM) which provides a framework for planning, designing, and managing urban water systems (GWP 2013). It is a flexible process that responds to change and enables stakeholders to predict the impacts of interventions (Bahri 2012; GWP 2013). IUWM includes environmental, economic, social, technical, and political aspects of water management. It brings together freshwater, wastewater, stormwater, and solid waste, and enables better management of water quantity and quality. IUWM calls for aligning urban development with basin management to ensure sustainable economic, social, and environmental relations along the urban–rural continuum. IUWM is implemented to cater for the rise in the world population (more than 7 billion), overcrowded informal settlements, rapid urbanisation, inadequate water infrastructure and climate change effects.

4.2.2 Integrated Urban Water Management Project: Marondera Case Study

The Integrated Urban Water Management (IUWM) project aims at providing the Municipality of Marondera with an innovative and integrated approach that will ensure sustainable management of water and sanitation for Marondera's population (AFDB 2022). The project will address the immediate Water Supply and Sanitation (WSS) needs, the development of an Integrated Urban Water Management Master Plan for the Municipality and enhance the capacity of the Government of Zimbabwe and key stakeholders to implement sustainable water and sanitation facilities (AFDB 2022). The IUWM contributes to improved water security through the adoption of a holistic approach. This involves designing water management as a system within a catchment, taking into account the entire water cycle in general, and in particular, that wastewater and faecal sludge are resources (AFDB 2022). The project total cost is 2,340,450 Euros.

4.2.3 Case of the Current WASH Situation in Zimbabwe Towns and Cities

In most Zimbabwean towns and cities, the rapid increase in urbanisation is coupled with a rapid increase in informal settlements in the urban dwellings. This has prompted a rise in demand for housing, water and sanitation services on the local authorities with more than 60% of the urban dwellers living in informal settlements. Most towns and cities are constantly under threat from water borne diseases such as cholera, typhoid and dysentery with more than 6000 cholera deaths reported in urban areas due to lack of potable water supply which has driven residents to shallow wells/ boreholes which tend to be contaminated by pit latrines in areas with high water table or during the rainy season (United Nations Habitat 2020).

Zimbabwe continues to receive assistance from the United Nations Habitat to facilitate the establishment of urban governance, housing regulations and policies, capacities in urban planning, water and sanitation in peri-urban settlements and city resilience strategies. In the period 2018–19, the UN Habitat implemented a Water and Sanitation project in Epworth which benefitted more than 15,000 residents. The project focussed on supporting the vulnerable in peri-urban areas. Those who benefitted included 70% of women headed households, 7000 youth and children, 80% of the elderly who were beneficiaries of reticulated toilets and 30% disabled who had challenges with access and use of pit latrines.

In rural areas, access to safe potable water and sanitation is still a challenge. More than 31.3% of the rural population is reported to be still practising open defaecation with more than 40% dysfunctional water points (ZIMSTAT and UNICEF 2019). The United Nations Children's Fund (UNICEF) is working with the Government of Zimbabwe and other development partners to improve equitable use of safe drinking water, sanitation and hygiene practices in rural areas. In 2015, the Rural WASH Information Management System (RWIMS) was established in seven out of ten provinces to enable management of WASH data; generation and sharing of information; stakeholder connection; and enhanced service delivery. To date, the RWIMS has managed to capture, manage and analyse WASH data generated at the sub-national level within one national system. According to UNICEF (2021), through the real time date generated by RWIMS, by December 2020 at least 5.4 million children had access to water and sanitation services in Zimbabwe.

In Chipinge, Ward 16 Kudzika, the World Food Programme in partnership with Goal Zimbabwe are implementing a project under the Food Assistance for Assets. A total of 6 solar powered boreholes were drilled to irrigate 12 hectares of land. The non-governmental organisation is also building double squat hole toilets, conducting trainings, and health harvest modules for the community. The communities are also encouraged to wash hands and dig rubbish pits as part of hygiene practises around the compound.

In recent years, the intervention of non-governmental organisations and improved funds from the Government of Zimbabwe has significantly improved the wash situation in urban centres. Funds have been availed to provide support for the supply of essential water treatment chemicals, rehabilitation of critical components of the water and sanitation system, institutional capacity development for water utilities; and strengthening of WASH policy and institutional frameworks. Areas where rehabilitation work was carried out include:

- Mhlanga water supply station in Plumtree, rehabilitation of 2 lift pumps,
- Bindura water supply station, rehabilitation of a chemical dosing station,
- Installation of a new water pump at Shurugwi water treatment plant,
- Rehabilitation of sedimentation tank at Chipinge water treatment plant, and
- Rehabilitation of a sludge pump station at Karoi water treatment plant.

In September 2018, the Government of Zimbabwe launched Vision 2030 which envisages transforming Zimbabwe into a prosperous and empowered upper middle income society by 2030. The vision takes into cognisance the United Nations' Sustainable development goals for the period 2016–2030. In the water and sanitation sector, the government vowed to prioritise the provision of adequate water and sanitation infrastructure to address the backlog and also meet the high demand of households and industries emanating from population and urbanisation growth. A master plan was developed to guide the realisation of water, sanitation, waste management deliverables and numerous investments are lined up in the water sector, benefitting dam construction for water storage beyond the economy's needs, and to include supply to some of the neighbouring countries. In the rural areas, borehole drilling programmes coupled with conveyance systems development are being undertaken, in partnership with the private sector and development partners (Government of Zimbabwe 2018).

Water, sanitation and hygiene has multiple vital economic linkages to key sectors such as agriculture, energy and mining. In the NDS 1 (2021–2025), the Government of Zimbabwe targets to increase access to potable water from 77.3 to 90% and the storage capacity from $15.423*10^6$ to $16.979*10^6$ mega litres by 2025. A National Dam Safety Plan was proposed with the objective of strengthening existing capacities for water resources management and further development of water resources to cater for existing and future demand as well as reducing hydrological and climatic vulnerability. To address the issues of sustainability of the water resources, a reinforcement in demand management was proposed. Moreover, Zimbabwe National Water Authority operations will be commercialised to ensure cost recovery when supplying raw and treated water for agricultural, household and industrial use. Zimbabwe Vision 2030 targets the acceleration of construction of major dams which significantly improves water supply situation for urban centres. These include the following:

i. ***Kunzvi-Musami Dam***

The project involves construction of a dam, conveyancing and treatment works to supply water to Harare, Chitungwiza, Ruwa and Norton residents and also provide irrigation water to surrounding communities. The site is approximately 67 km from Harare on the confluence of Nyaguwe and Nora rivers in Goromonzi District. The anticipated capacity is 250,000 m^3/day, with an estimated project cost of US$865 million funded through Public Private Partnership initiatives. Dam location is upstream therefore; treatment costs savings are anticipated in comparison to downstream Lake Chivero. In the interim, the Harare City Council rehabilitated and upgraded the Morton Jaffray water purification plant, to alleviate water woes in the city.

ii. ***Gwayi-Shangani Dam***

The dam is located approximately 6 km downstream of the confluence of the Gwayi and Shangani Rivers. Upon completion, it will be the 3rd largest inland water body in Zimbabwe with a storage capacity of 634 million m^3. The main contractors are

Zimbabwe National Water Authority (ZINWA) and China International Water and Electric Corporation (CWE). The project is part of the greater Matabeleland Zambezi Water Project, a long-term solution to the perennial water challenges in Matabeleland North, South, and Bulawayo provinces. Furthermore, a 260-km-long water pipeline from Gwayi-Shangani to Bulawayo is under construction. To date, boreholes were commissioned in the Epping Forest, drawing at least 10,000 m^3/day from Nyamandlovu aquifer, for Bulawayo residents.

According to Government of Zimbabwe (2020), sanitation will improve from 70.22 to 77.32% in both urban and rural areas with particular attention to reducing open defecation in rural areas, by the year 2025. The key strategies to improved sanitation services are: expanding on-going programmes that are targeting rehabilitation of the existing urban and rural network of sanitation facilities; increasing the access to improved sanitation; implementing institutional reforms that will strengthen coordination and implementation of sanitation programmes and expand financial support for the programme; and reinforcing routine maintenance to expand the on-going hygiene education programmes for urban and rural communities.

WASH services in Zimbabwe are way below pronounced government commitments and still lags behind by far in comparison to its regional counterparts. Zimbabwe has an access to improved water sources of 20% whereas Botswana, 96%; South Africa, 93%; Lesotho, 82%; and Swaziland, 72%. In Zimbabwe, public investment in WASH remains biased towards urban areas, with most allocations being towards the Small-Town Water Supply projects and Urban WASH despite the evidence suggesting that rural areas, face significant challenges with regards to access to improved WASH. Table 2 presents the WASH indicators in Zimbabwe based on the UNICEF Zimbabwe (2018) report.

To date, tracking of WASH resources is a challenge since they are managed by various Ministries and departments. Better coordination within the WASH sector is essential to facilitate enhanced reporting and accountability for WASH resources moreover creating avenues for better resource mobilisation for the sector (UNICEF Zimbabwe 2018). Improvement in coordination will go a long way in increasing spending on critical issues such as community mobilisation and hygiene promotion.

Table 2 WASH indicators in Zimbabwe

Indicator (%)	Total	Urban	Rural	Best province	Worst province
Improved source of drinking water	76.1	67.5	98.4	96.9 (Bulawayo)	64.3 (Masvingo)
Improved drinking water sources and improved sanitation	29.7	23.1	46.9	54.7 (Bulawayo)	17.5 (Mat North)
Open defecation	31.7	43.5	1.1	0.1 (Harare)	69.6 (Mat North)
% of children whose stool was last disposed of safely	57	46.9	86	84.8 (Harare)	19.8 (Mat North)

Source UNICEF Zimbabwe (2018)

5 Smart Settlement Models

There is consensus among researchers that "smart settlements" are generally made up of a mix of smart economy, smart energy, smart infrastructure (hard infrastructure and soft infrastructure), smart environment, smart mobility, smart education, smart health and smart governance. Smart cities are mainly centred on 5G, high speed fibre, Internet of things IOT, information and communication technologies (ICT) and data analytics (Kummitha and Crutzen 2017). Most smart cities rely on ICT to ensure efficient management of resources which ultimately conserves the environment. This has numerous benefits for the general populace who get to enjoy the public benefits like transportation. Smart cities are thus, digitally driven cities which have efficient transportation, waste management, energy provision, water supply, health and educational facilities. Frost and Sullivan Group (2020), posit that these technologies in smart cities will generate business opportunities worth US$ 2.46 trillion by the year 2025. Zion Market Research (2019), estimates that IoT will grow by 22.6% in 2025. It is projected that by 2026, 87% of IoT devices will use smart metering for water and energy supply (Carlow et al. 2022). Blockchain technology which has been implemented in Estonia promotes transparency and helps fight corruption (Junestrand et al. 2018) is another technology which is projected to grow beyond 20 billion dollars by 2024 (Transparency Market Research, 2018). Geospatial technology can also be used to make environments more sustainable and this technology is projected to grow to 134 billion dollars by the year 2025 (Grand view Research 2019). Africa in general and Zimbabwe in particular should thus embrace these technologies and get involved in the business of the smart city technologies to earn revenue.

The quest to have smart cities in Africa as outlined in the continent's agenda 2063 (Slavova and Okwechime 2016) is hampered by economic and political challenges in most regions (Dos Santos and Mota 2019). The smart cities planned in Africa include Hope City in Accra, Ghana; Luanda Satellite City and Nova Vida in Angola; Kigamboni City close to Dar-e-Salaam, Tanzania; and Cite le Fleuve in the Democratic Republic of Congo (Watson 2014). The gap between the rich and the poor continues to widen in Zimbabwe just as in most parts of Africa. This is a threat to the glitz of new cities as the poor will inevitably clog the streets (Watson 2014). In Africa, most plans for smart cities or Eco-Cities are extensions or adjacent developments to already existing cities without clear plans of integration with the existing. Establishment of a smart city in Kigali, Rwanda is a perfect example of a smart city which was clearly planned to replace an existing one however, problems of exclusion of the poor blight the city. The demand by Johannesburg Municipality to have 5000 housing units for the poor in Modderfontein New City, tipped to be the capital of Africa, led to the collapse of the project as project financiers and real estate companies were mainly targeting the elite and profits (Reboredo and Brill 2021). There is however an opportunity for adoption of smart cities in Zimbabwe. Zimbabwe has the advantage of peace and stability which can overcome economic volatility. Also despite the challenges, trajectories show that Africa will be a dominant economic giant in years to come thus, real-estate companies are actively seeking to invest in new smart cities in Africa. Zimbabwe can also learn from South Africa as they have planned more smart cities like Mooikloof, Lanseria and Durban Aerotropolis despite the failure of Modderfontein. The ongoing construction of the Eko Atlantic, a coastal city next to Lagos in Nigeria is another example that smart city models can be fully funded and achieved in Africa. The smart city of Konza, Kenya gives further hope that Africa can replicate the smart cities in Shanghai, Dubai and Singapore. Smart city models are indeed the way forward for Africa since they have real time data gathering which can be managed in an integrated data management system with endless possibilities for efficiency. Zimbabwe is on the right track as evidenced by the use of geospatial technology (Magure 2022) in the development of the digital Zimbabwe wetlands masterplan (Environmental Management Agency 2022) and in the use of GIS in non-revenue water monitoring (Urban Councils Association of Zimbabwe 2021). The establishment of the Zimbabwe National Geospatial and Space agency (ZINGSA) in 2018 is also further testimony of the government's commitment to embracing smart city technologies. The willingness by government to embrace the smart city technologies is a major step towards achievement of the NUA agenda in Zimbabwe. The government has already initiated plans for smart settlements in Tugwi Mukosi, New City (Mount Hampden) among other new smart cities being mooted. Singapore tops the list of urban mobility infrastructure and investments above cities like London and Shanghai among others (Mazareanu 2022). Zimbabwe should also aim for these high standards set by smart cities in developed countries while learning from the mistakes made in other smart city developments in Africa.

Zimbabwe is currently working on master plans for two smart cities namely Tugwi Mukosi and New City Mount Hampden Master Plan. Tugwi Mukosi city is proposed to be built in the environs of the Tugwi Mukosi dam, the largest inland dam within Zimbabwe. It is a concrete faced rock fill dam constructed at the confluence of the Tugwi and Mukosi Rivers in Masvingo Province. Construction of the dam was completed in 2017 and in line with the government's vision 2030, a master plan for development around the dam is being developed which includes a smart city. In the same vein, there are plans for a smart and sustainable new city on the 18,000 ha of land centred around the new parliament building in Mount Hampden. The setting of the new city has both traditional and historical connotations as oral tradition states that, it was the original site identified for the location of Harare by the colonial regime. In addition to key government ministries, the new city will have all other infrastructural amenities associated with a modern eco-city. The plan is to produce a city which will be the envy of the world by merging societal needs with economics and environmental preservation. This will be achieved by proposing policies and standards for green buildings and smart settlements. The plans by the Zimbabwean government are in tandem with the NUA and show the possibility of fulfilment of the agenda.

6 Conclusions and Recommendations

The study sought to focus on the contribution of construction and civil engineering to the New Urban Agenda (NUA) and sustainability in Zimbabwe. It concluded that, there is a vital role of engineering in achieving sustainability through the role it plays in shaping the built environment. The study provided engineering innovations that are shaping the country, especially emerging technologies, which are crucial for addressing the pressing challenges facing humankind and the planet. It is therefore crucial for the profession to continuously transform its education and capacity-building at the dawn of the Fourth Industrial Revolution as this will enable engineers to tackle the challenges ahead. By presenting case studies, the paper reveals the reasons engineering is critical for sustainability and how the role of engineers is important in addressing basic human needs such as providing housing and road infrastructure, supplying clean water and energy, responding to natural disasters and constructing resilient infrastructure.

The study makes the following recommendations:

- For a successful implementation of NUA in Zimbabwe, there should be a participatory approach in terms of urban planning and stakeholder engagement. The government and stakeholders should be fully committed to the implementation of these goals, through financial backup, zero tolerance to corruption and introduction of stiffer penalties to land barons who parcel out land illegally.
- Industry should embrace the sustainability agenda and the teething pains which might come with it. This will involve adoption of new construction methods, materials, standards and by-laws, and training staff on new methods and legal requirements. The agenda will place a need for industry to create more employment and investment in communities.
- Unity for fostering engineering development through investment and cooperation in the country so as to make engineering a true enabler, equaliser and accelerator to deliver on the NUA.
- Review current national engineering education benchmarks for graduate attributes and professional competencies to ensure that they meet the requirements of current and future employers, and incorporate the values and principles of sustainability, diversity and inclusion, and ethical engineering practice.

References

AFDB (2022) Zimbabwe-integrated urban water management project. https://projectsportal.afdb.org/dataportal/VProject/show/P-ZW-EAZ-002. Updated 31 Jan 2022

Bahri A (2012) TIC background papers No. 16, integrated urban water management. Global water partnership technical committee (TEC). ISBN 978-91-85321-87-2. www.gwp.org

Carlow S, Bonte D, Mavrakis D (2022) ABI research. Allied Business Intelligence, Inc. https://www.abiresearch.com/

Dos Santos ML, Mota M (2019) Toward sustainable and smart cities in Africa: a review and challenges. In: Guedes M, Cantuaria G (eds) Bioclimatic architecture in warm climates. Springer, Cham. https://doi.org/10.1007/978-3-030-12036-8_10
Ebohon OJ, Rwelamila PD (2001) Sustainable construction in Sub-Saharan Africa: relevance, rhetoric, and the reality. Agenda 21:1–16
Environmental Management Agency (2022) Zimbabwe national wetlands masterplan, Harare, government printers
Frost and Sullivan Group (2020) Smart cities to create business opportunities worth $2.46 trillion by 2025. https://www.frost.com/news/press-releases/smart-cities-to-create-business-opportunities-worth-2-46-trillion-by-2025-says-frost-sullivan/
Garschagen M, Porter L, Satterthwaite D, Fraser A, Horne R, Nolan M, Solecki, W, Friedman E, Dellas E, Schreiber F (2018) the new urban agenda: from vision to policy and action/will the new urban agenda have any positive influence on governments and international agencies? / Informality in the new urban agenda: from the aspirational policies of integration to a politics of constructive engagement/growing up or growing despair? Prospects for multi-sector progression city sustainability under the NUA/approaching risk and hazards in the new urban agenda: a commentary/follow-up and review of the new urban agenda. Plann Theor Pract 19(1):117–137. https://doi.org/10.1080/14649357.2018.1412678
Government of Zimbabwe (2018) Vision 2030. In: Towards a prosperous and empowered upper middle society by 2030. Government of Zimbabwe, Harare
Government of Zimbabwe (2020) National development strategy 1. Government of Zimbabwe, Harare
Grand View Research, Inc. (2019) Geospatial analytics market size worth $134.48 billion by 2025: Grand View Research, Inc. https://www.prnewswire.com/news-releases/geospatial-analytics-market-size-worth-134-48-billion-by-2025-grand-view-research-inc-300892101.html
GWP (2013) Policy brief, integrated urban water management (IUWM), towards diversification and sustainability. www.gwp.org
Junestrand S, Javier A, Gildo S (2018) Blockchain4Cities: investigación sobre blockchain aplicado a la gobernanza de las ciudades inteligentes. E-smart city. https://www.esmartcity.es/comunicaciones/comunicacion-blockchain4cities-investigacion-blockchain-aplicado-gobernanza-ciudades-inteligentes
Kummitha RKR, Crutzen N (2017) How do we understand smart cities? An evolutionary perspective. Cities 67:43–52
Magure M (2022) Wetland fragmentation and hydrologic implications in Zimbabwe's Highveld. Thesis for MSc in Integrated water Resources Management at the University of Zimbabwe, Harare
Mapangire B, Shumba S, Mushiri T (2019) Design of an integrated fruit and vegetable processing factory located at Juru Service Centre, Mashonaland East, Zimbabwe. In: 10th international conference on structural engineering and construction management (ICSECM) 12–14th December 2019. Earl's Regency Hotel, Kandy, Sri Lanka, pp 309–325, ISBN 978-955-589-274-2 ICSECM2019-48
Mazareanu E (2022) Urban mobility readiness index 2022 Report. https://www.oliverwymanforum.com/content/dam/oliver-wyman/ow-forum/template-scripts/urban-mobility-index/PDF/Mobility-Index-Report.pdf
Mukura TE, Shumba S, Vassileva LD (2018) Construction of low volume roads using labour-based technology: case study of Mutare Zimbabwe. American Society for Testing and Materials (ASTM) Advances in Civil Engineering Materials. Special Issue on Advances in Pavement Technologies in the Emerging Economics. ACEM-2017-0038. 7(2):272–290. ISSN 2379–1357. Available online at www.astm.org. https://doi.org/10.1520/ACEM20170038
Ofori G, Hindle R, Hugo F (1996) Improving the construction industry of South Africa: a strategy. Habitat Int 20(2):203–220
Philip T, Shumba S, Vassileva LD, Cheure N (2017) Analysis and redesign of Simon Mazorodze and Chitungwiza Road Intersection in Harare, Zimbabwe. In: Proceedings of European alliance for

innovation (EAI) international conference for research, innovation and development for Africa held from 20–21 June 2017 in Victoria Falls. Published in European Union Digital Library, pp 52–64. http://eudl.eu/doi/10.4108/eai.20-6-2017.2270747

Reboredo R, Brill F (2021) What a failed Johannesburg project tells us about mega cities in Africa. https://theconversation.com/what-a-failed-johannesburg-project-tells-us-about-mega-cities-in-africa-112420. Copyright © 2010–2021, The Conversation Africa, Inc

Slavova M, Okwechime E (2016) African smart cities strategies for agenda 2063. Afr J Manag 2(2):210–229

Transparency Market Research (2018) Blockchain technology market to gain revenue worth US$20 bn by 2024. https://www.prnewswire.com/news-releases/blockchain-technology-market-to-gain-revenue-worth-us20-bn-by-2024---tmr-686428221.html

UNICEF Zimbabwe (2018) Water, sanitation and hygiene (WASH) budget brief. UNICEF, Harare

United Nations Children's Fund (2021) Accelerating the delivery of water, sanitation and hygiene services, with rural communities using real-time data and rapid response to service disruptions: good practices and lessons learned in national systems strengthening using mobile technology in Zimbabwe. UNICEF, New York

United Nations Habitat (2020) UN-habitat Sub-Saharan Africa Atlas. United Nations, Nairobi, Kenya

Urban Councils Association of Zimbabwe (UCAZ) (2021) Service level bench marking. https://ucaz.org.zw/service-level-benchmarking-slb/

Valencia SC, Simon D, Croese S, Nordqvist J, Oloko M, Sharma T, Buck NT, Versace I (2019) Adapting the sustainable development goals and the new urban agenda to the city level: initial reflections from a comparative research project. Int J Urban Sustain Develop 11(1):4–23. ISSN 1946-3138. https://doi.org/10.1080/19463138.2019.1573172

Watson V (2014) African urban fantasies: dreams or nightmares? Environ Urban 26(1):215–231

Watson V (2016) Locating planning in the new urban agenda of the urban sustainable development goal. Plan Theory 15(4):435–448

Zimbabwe National Statistics Agency and United Nations Children's Fund (2019) Zimbabwe multiple indicator cluster survey 2019, survey findings report, ZIMSTAT and UNICEF, Harare, Zimbabwe

Zion Market Research (2019) IoT in smart cities market—global industry analysis. In: IoT in smart cities market by component (solution and service), by application (lighting, traffic, utilities, public safety, environmental monitoring, and others), and by end-user (information and technology, telecommunication, government, automation, energy, and others): global industry perspective, comprehensive analysis, and forecast, 2018–2025

Zuo J, Zhao ZY (2014) Green building research–current status and future agenda: a review. Renew Sustain Energy Rev 30:271–281

The Role of Architecture in Implementing the New Urban Agenda

Stephen Zavirima, Brilliant Mavhima, and Charles Chavunduka

Abstract The major contribution of architecture to urbanism is in the provision of a framework for housing, being a set of infrastructure and services that cater for human needs. In Zimbabwe, the lack of collaboration between architects and other built environment professions has reduced the capacity of architecture to contribute meaningfully in steering the New Urban Agenda. This chapter explores opportunities in education and practice for enhancing architects' contribution in realising the New Urban Agenda in Zimbabwe. It is based on a desk review and structured interviews with architects in academia and practising institutions, with the data subjected to textual analysis to derive meaning and interpretation. The results show that the architecture profession in Zimbabwe does not provide adequate scope for collaborating with other built environment professionals thereby sub-optimising architecture contribution to the achievement of the Sustainable Development Goals. The role of the architect has been confined to the design and supervision of building construction at the expense of other important duties like the carrying out of feasibility studies, environmental impact assessment, economic analysis and project management. Enhancement of the contribution of architecture to the achievement of the Sustainable Development Goals requires legal reform towards its teaching and practice as an interdisciplinary profession. The interdisciplinary approach will ensure that the creation of new cities, neighbourhoods and housing areas will not be left to discrete professions but will be done through collaborative and interdisciplinary efforts. The interdisciplinary approach will deliver more affordable, inclusive and sustainable housing thereby driving communities towards the achievement of Sustainable Development Goals.

Keywords Architecture · Sustainable · Inclusive · Interdisciplinary · Collaborative · New urban agenda

---

S. Zavirima (✉) · B. Mavhima · C. Chavunduka
Department of Architecture and Real Estate, University of Zimbabwe, Box MP 167, Mt. Pleasant, Harare, Zimbabwe
e-mail: szavirima@eng.uz.ac.zw

C. Chavunduka and I. Chirisa (eds.), *New Urban Agenda in Zimbabwe*,
Advances in 21st Century Human Settlements,
https://doi.org/10.1007/978-981-97-3199-2_7

1 Introduction

The focus of the New Urban Agenda (NUA) is mainly inclined towards the renewal of the global commitment to sustainable urbanisation, building on the initiatives adopted at Habitat III held in Istanbul, Turkey, in 1996. The NUA sets a new global standard for sustainable urban development which helps in rethinking how to plan, manage and live in cities. It is a roadmap for building cities which promote social wellbeing while protecting the environment and acting as engines of prosperity. It further helps in efforts to meet the challenges of rapid urban development and climate change (Caprotti et al. 2017). Commitments made in the NUA are: provision of basic services for all citizens, ensuring that all citizens have access to equal opportunities and face no discrimination, promoting measures to support cleaner cities, reducing greenhouse gas emissions to address the issues of climate change and improving connectivity and support innovative and green initiatives as well as promoting safe, accessible and green public spaces (United Nations 2017). According to Hegemann et al. (2008) without the architects' insight, other issues may go unaddressed. The issues include gentrification, inclusive house designs among others. Global estimates indicate that by the year 2050, 70% of the people will be living in urban areas (Zajda 2021). Cities and other urban settlements become risky places in the context of employment, climate change and poverty among other issues (Granberg and Glover 2021). Organisations like the American Institute of Architects and the Royal Institute of British Architects have taken interest in the NUA, tailoring the architectural philosophy into what is termed the Communication, Advocacy and Policy of New Urban Agenda (CAPNUA). The approach indicates a global reorientation of the architectural field.

In sub-Saharan Africa, urbanisation problems are rapidly increasing. Such problems include inequality, absolute poverty, poor housing, congestion, poor environmental conditions, absence of public transport as well as a run-down infrastructure (Ali 2021). The situation has paved way for the development of the New Urban Agenda. The African Agenda 2063 coupled with the NUA opens avenues for professionals to manage the African urban problems. Such opportunity comes in Africa where local architecture is being forsaken for seemingly modern imported architecture (Angelopulo 2021). In 2008, Chinese investors and workers erected a new Angolan town called Kilamba, covering 5000 ha, 30 km outside the capital Luanda. The settlement was developed with identical high-rises, surrounded by barren land devoid of anything natural such as trees. It gives a solid example of a culturally alien approach, implemented with no regard for actual needs or indigenous traditions. The settlements were designed without considering uses like fields and gardens, streets and markets, entertainment venues, mosques and churches (Highet 2013), a situation which makes it difficult for the development of sustainable and \community-friendly settlements as indicated in the NUA.

Zimbabwe has been developing in a planned but incremental way. Statistics indicate that the country is now 32.24% urban (Chigudu 2021). The rapid urbanisation places the NUA into perspective with the idea being of taming the urban settlements.

The nation has engaged in various policy options with the new National Human Settlements Policy being the most recent attempt. The document gives avenues for professional integration towards the development of the country' urban areas. This attempt came following the existence of a disintegrated approach to city planning where built environment specialists just focused on one's line and avoided any forms of interdisciplinary approach (Matamanda et al. 2021).

The chapter addresses two questions. Firstly, how has architecture enabled the achievement of the NUA in Zimbabwe? Secondly, how can architecture be shaped to be more effective in enabling the NUA? While architects are often viewed as micro-environmentalists, their role in the built environment is very significant. The main role of architecture is to solve complex issues of human life ranging from culture, religion, politics, emotions, status, communication, health, environment and those beyond the cosmos through the means of design. Architecture plays an important role in the planning of cities and the NUA. The next sections present the conceptual framework and literature review. The methodology that was used in the study is then explained. This is followed by presentation and discussion of findings. The conclusion emphasises the need for a built environment professional integration in driving the NUA.

2 Theoretical Framework

Architecture is defined as the designation of space (Ulusoy and Kuyrukcu 2012). It is the art of designing structures required for making the lives of people easy and enabling them to engage in and maintain such activities as sheltering, resting, working and entertaining. In an urban setting, the resultant physical structures complimented by external spaces and artefacts under the control of planning and zoning rules by extension have a direct impact on the environment by way of dictating its character and uniqueness and how it is perceived and interpreted visually and physically. The architecture discipline is both an art and a science that includes the design and construction of buildings in the physical environment. The process of designing buildings plays a significant role in shaping communities. This can be seen through understanding the concept of architectural determinism. Architectural determinism borrows principles from environmental determinism in the sense that like climate the architectural design of an area determines the behaviour of the community (Freilich 1967). The relationships and social networks of an area are, therefore, influenced by the architectural form of the area. However, these ideas have been counteracted by environmental possibilism theory which refutes the claim that architectural designs necessarily determine the behaviour of inhabitants of an area. The belief that architectural design solutions would change behaviour predictably and positively turned out not to be true (Tafuri 2006). It had been the claim of architects for decades that their designs would reshape society through the power of their art, which remains a lovely if unsubstantiated notion.

Another theoretical perspective to architecture distinguishes between the traditional and contemporary roles of an architect (Jarman et al. 2018). Traditionally, an architect was seen as a designer, draftsman, project manager and general overseer of a project with overall liability of the project (Koch and Worndl 2001). In contemporary times, the architect is seen as part of a multifaceted but well-structured project team. The project team has equal responsibility and liability as opposed to discrete responsibilities and liabilities. The contemporary approach encourages cross-pollination of ideas and better solutions. Michael Sorkin, an American architect, listed 250 things that architects should understand which range from social life, health to technical issues in order for them to produce sustainable buildings (COR 2022). His list and explanation are prolific and eye-opening in that, it shows that architecture cannot be left under one profession. In Zimbabwe, the Architects Act is based on the traditional approach that sees architecture as a discrete profession.

3 Methodology

The chapter draws from a study mainly based on a desktop research, literature review, complemented by structured interviews with five architects. Reviewed documents include the New Urban Agenda, Agenda 2063, Zimbabwe Vision 2030, the National Development Strategy 1, the Architects Act (Chapter 27:01) and building by-laws. Face-to-face structured interviews were done where architects were asked about the contribution of Zimbabwean architecture education and practice to the NUA. They were also asked about what architecture education and practice need to address in order to effectively contribute to the NUA. Gathered data was subjected to thematic analysis. The analysis procedure involved data coding, searching for themes and recognising relationships and refining themes (Saunders et al. 2019).

4 Role of Architecture in the Development of Human Settlements

Architects have discussed the potential influence of architecture since the 1400s. In the 1400s, Italian Renaissance-era architect Leon Battista Alberti claimed that balanced classical forms were so influential that they would compel aggressive invaders to down their arms and become civilians. Furthermore, the US architect Frank Lloyd Wright believed that when done right, architecture would save the USA from corruption and turn people back to "wholesome endeavours". The Swiss-born French architect Le Corbusier claimed that the power of his designs for his Villa Savoye would heal the sick. While all the claims seem to be intriguing, they have been criticised for not being entirely true (Horne 2011). However, the claims are indicative of the potential influence of architecture in the implementation of the NUA.

The role of architecture in community development was criticised severely in America. A case in point was the demolition of the infamous Pruitt-Igoe urban housing complex in St Louis in the USA. George Hellmuth, Minoru Yamasaki and Joseph Leinweber designed the complex with the idea that "community gathering spaces and safe, enclosed play yards" would provide a happy and prosperous living environment. The power of architecture was shown to be a double-edged sword, that provided a hotspot for crime and poverty, as such, it was finally demolished in the 1970s, with architects raising their hands and for once admitting they got it wrong.

Evidence of the impacts of architecture has been seen on the history and position of architectural products on human behaviour. Research indicates that large grey landscapes and boring buildings have been found to cause higher levels of stress (Gold 2013). Without variety and stimulation, the human mind gets confused and is reminded of just how far out of its natural habitat it is. Boring and uninspiring architecture could cause real harm to the community. Conversely, a 2009 paper by Murray Krelstein, MD in the Psychiatric Times entitled "Have You Ever Felt Awe and Wonder?" suggests that awe reduces the prevalence and severity of mood disorders, meaning that beautiful and awe-inspiring architecture has the potential to create a calmer, happier environment.

Traditionally, there were no professional architects, urban planners or engineers but cities were constructed (Rudofsky 1964). However, many cities which were constructed without the involvement of professionals are either in ruins, in need of massive rehabilitation or dissolved into oblivion. Though there are several reasons why many ancient cities disappeared, it is necessary to understand that they were not professionally planned and were not sustainable.

Architecture is an all-encompassing discipline which does not exist as an object but as a system of functions and processes to serve human needs. It does not only deal with the micro-organisation of functions but starts from the macro-holistic point of view. In order for an architect to produce a piece of architecture, it requires a knowledge pool of resources, from different facets of life. The architect's role is to organise social, political, economic and natural functions in the most efficient and effective way in order to satisfy human needs.

People view architects as creators of beautiful buildings, although the creation of drawings is a task involving substantial input from lower cadres (Perez-Gomez 1982; Hind et al. 2022). In the production of drawings, there is substantial input from technicians, draftsmen and architectural technologists. The main role of architects is to provide creative, well-thought and planned structures that consider political, social, economic and natural contexts (Brussat 2020). Architects organise functions in space taking into consideration all facets of human life; from politics, economics, psychology, biology and chemistry to physics. The huge task of coming up with a functional system of multifaceted elements means that if architecture is left in the hands of one profession, many issues will be left unresolved and can become a time bomb (Barkman 2021).

In Africa, architecture is playing a role of addressing the challenges of rapid urbanisation, informality and climate change (Horne 2011). Currently, only 43% of the African population reside in urban areas but the rapid rate of urbanisation in the continent, which is only second to Asia, will lead to urban spaces accommodating 61.6% of the continent's population by 2050 (Matamanda et al. 2021, p. 2). There is proliferation of informal settlements in the Global South, with 61.7% of the urban population in Africa living in informal settlements, a figure that has been on the rise (Dovey et al. 2020; UN-Habitat 2014). Climate change has shown in recurrent heatwaves, droughts and water stress in Zimbabwean and Mozambiquan towns and cities. The increasing challenge of climate change requires new research, knowledge and experience in multiple areas, such as disaster risk management, resilient infrastructure and disaster preparedness. Higher education research and teaching of architecture have to accommodate sustainability and resilience and have academics and practitioners re-orientate their approaches. Additionally, Africans need to reclaim ownership of their heritage to inform local building approaches. Africa's sustainable architecture movement is already reducing costs, by learning from traditional building methods, conserving energy and promoting sympathetic reuse of buildings. Experimentation with sandbag houses has proved to be a strong, safe and cheap method of delivering affordable housing with low-maintenance materials, using the collective labour of residents.

While there is still no definitive answer to how architecture can influence society, it is still widely understood and accepted that architecture will always serve more than simply a physical functional purpose. The New Urban Agenda highlights linkages between sustainable urbanisation and job creation, livelihood opportunities and improved quality of life and insists on incorporation of all these facets in every urban development or renewal policy and strategy. The literature highlights promotion measures that support cleaner cities, reducing greenhouse gas emissions to address issues of climate change, improving connectivity and supporting innovative and green initiatives as well as promoting safe, accessible and green public spaces. There is need to understand the influence of Zimbabwean architecture in the NUA and place it within the wider global context.

5 Results and Discussion

Global trends indicate the significance of the field of architecture in the development of human settlements (Neuman 2021). In Zimbabwe, the role of an architect is defined and regulated by the Architects Act (Chapter 27:01). According to the Act, an architect is one who is registered by the Architects Council of Zimbabwe (ACZ). Section 39b of the Architects Act states that any person who is not registered with the ACZ shall neither do the work nor carry the title of an architect. Part 1, Section 2 of the Act describes two main functions of an architect. The functions of an architect are to design buildings, make additions and to supervise the construction or addition to buildings. The role of an architect as described by the Act is centred mainly on the creation of an object, and there is no regard to other important duties of architects like feasibility studies, environmental impact assessment, economic analysis and project management. The statutory definition of an architect agrees with the description of architecture as the art, science or practice of designing and building structures for human use that was given by one of the interviewed Zimbabwean architects.

The planning of settlements in Zimbabwe is a process that involves the engagement of different professionals at different times. The main professionals involved in the development of settlements are land surveyors, spatial planners, architects, quantity surveyors and civil engineers. Land surveyors carry out topographical surveys and produce a base map. The spatial planner uses the base map in the preparation of a settlement layout plan which is passed on to the civil engineer for road, water, sewer and other infrastructure designs. The layout plan is given to the land surveyor for the demarcation of plots and the production of a general plan. After that, the architect designs buildings for individual or several contiguous stands on the general plan. The current settlement planning process indicates the peripheral nature of architectural practice in the development of settlements. The settlement planning process in Zimbabwe is a heteronomous practice which is reliant on numerous professions that are discrete in their operations. The status quo of architecture in Zimbabwe has fallen behind global trends in the development of the urban design profession which approaches architectural practice from macro, meso and micro-scales (Dolman 2021). Elsewhere, urban design and planning practice has become an interdisciplinary process that brings together spatial planners, urban design, landscape architects, environmentalists and civil engineers to work on one project such as Hammarby Sjostad in Stockholm, Sweden. Hammarby Sjostad is a new environmental and ecological city district. This approach has brought architects to the centre of the NUA as they deliver on multiple human needs.

The various professionals that work with architects are governed by different laws and professional bodies. Although the roles are interlinked, the Architects Act provides no room for collaboration as such different professionals work separately. Section 37 (2b) of the Architects Act on improper and disgraceful conduct states that:

> Notwithstanding anything to the contrary contained in this Act, by-laws referred to in paragraph (a) of subsection (1) may define that – any partnership relating to the work of an Architect; between a registered person and a person who is not a registered person or a member of a class of persons specified for purposes of this provision shall constitute improper or disgraceful conduct.

This subsection does not allow an architect to partner with anyone who is not a registered architect and this discourages interdisciplinary collaboration. A key informant felt that the idea that survey should be left to land surveyors, town planning to spatial planners, and structures to engineers and architects, is not suitable for the NUA because the power of collaboration will be lost.

The current situation of architecture in Zimbabwe lags behind trends across the world (South African Institute of Architects 2020). A survey of the architectural profession in the Commonwealth found that in order for a country to meet the demands for 2030 Agenda for Sustainable Development, it requires at least one architect per twenty thousand people (Jarman et al. 2018). Going by the findings of the survey, Zimbabwe requires a minimum of 816 architects for it to be able to successfully provide architectural services for its population and fulfil the 2030 Agenda for Sustainable Development. However, in Zimbabwe, only 56 architects are registered with the Architects Council of Zimbabwe. In South Africa, 11,551 professionals are registered to practice as architects. A survey conducted by the authors shows that non-registered academically qualified and non-academically qualified persons are practising as architects in contravention of the Architects Act. There is a wave of uncontrolled architectural activities that have spelt disaster on the economy and the public (Chombo and Youm 2021). Uninformed designs are proliferating with detrimental effects on the urban fabric, which has resulted in demolitions and loss of property and wealth. During Operation Murambatsvina (Clean the Filth), an estimated 700,000 people in cities across the country lost their homes that were deemed to be illegal structures (Tibaijuka 2005). More recently, Mtomba (2015) found that there were approximately 6000 illegal structures in over 20 settlements in Harare.

In addition to constraints posed by the Architects Act, the 1977 Model Building By-Laws which have been adopted by most local authorities were found to be outdated and a source of delays in the implementation of construction projects. The main structural materials specified in the by-laws are concrete, steel and timber. The use of other building materials requires the architect to prove their suitability to local authorities through an often tedium, costly and time-consuming process. The procedures slow down the rate of adoption of new building materials such as structural glass which may be as strong as steel.

Architectural education in Zimbabwe has contributed immensely to the NUA by preparing students to contribute to the built environment once they have completed studies. In order to manipulate the design and construction process, with the view of achieving the best possible results, particular skills are required to be learnt at higher education levels. In Zimbabwe, the first school of architectural training was established in 2000 at the National University of Science and Technology (NUST). The University of Zimbabwe followed suit by introducing architectural programmes in 2020. Architectural education makes students aware of historical movements in architecture as design is never static but always shapes itself to challenges that each generation faces and evolves with technological change. In history, it is taught that from around 900 BC, the Greeks in the mainland (the Peloponesse and the Aegon islands) developed Greek architecture which was mainly made of a thick base, doric, corithian or ionic columns, and a pitched roof. This was partly a result of available materials and technology. Various architectural movements followed until the work of Le Corbusier where buildings were lighter and supported on pilotis, as depicted in the villa savoir.

In Zimbabwe, architectural history is taught on how the Great Zimbabwe was built using available materials (mostly stone). Lessons have also been drawn from modern architecture as seen in Mick Pearce's masterpiece, the Eastgate Shopping Mall in Harare with its advanced cooling system whose concept is derived from an anthill. The building is designed to be naturally ventilated thereby supporting the green concept. The Reserve Bank of Zimbabwe building is ensconced in cultural motifs and is shaped like the Great Zimbabwe central tower with chevron patterns and the Zimbabwe bird to emphasise Zimbabwean identity.

Despite the critique of architectural "silos" approach (Barkman 2021), in 2021, the smart city project some 18 kms to the north-west of Harare brought together spatial planners, land surveyors, geographers, transportation planners, historians, anthropologists, sociologists, economists, land economists, realtors, business administrators and architects as a multidisciplinary team in its planning. However, it should be noted that this was an isolated event as, in general, there has been no collaboration among the built environment professions. The term "silo" in this context refers to an information system that is incapable of co-operation with others (De Waal et al. 2019). The smart city features sustainability concepts centred on green architecture, use of renewable energy and nature-based solutions to fostering resilient urban communities. The New Urban Agenda is one that should involve multiple disciplines but one of the constraints has been the lack of a coordinating institutional body that brings the built environment practices together. In the absence of a coordinating body, institutional collaboration has been ad hoc.

6 Recommendations and Conclusion

The concern with rapid urbanisation, climate change and informality calls for responsive architectural approaches. These concerns are driving a renewed interest in the use of local building materials and adaptation of building traditions to changing climatic conditions. Green architecture and participation of local communities in the building process using affordable and resilient local materials are the way to sustainable urban development.

In addressing these built environment problems, the Architects Act (Chapter 27:01) should be amended in order to allow the architectural profession to be collaborative so that it can harness ideas that can be obtained from various professionals and experiences in order for architecture to effectively contribute to the NUA in Zimbabwe. Such an approach would mean that architecture would not involve architects only as provided by the Architects Act (Chapter 27:01) Part 1 Section 2 but would encompass other fields that affect human life. The role of the architectural profession should change from a mere design of adorable artefacts to a multidiscipline that is multifaceted capable of providing sustainable housing. In architectural parlance, housing does not refer to dwelling units, but a set of infrastructure and services that provide for human needs (Mishra 1998; Desmond and Gershenson 2016; Richards 1979; Govender et al. 2011; Sadler et al. 2021). Proper housing should provide dwellings, places of work, places of rest, health and sanitary infrastructure; food, water, clothing and entertainment. The NUA requires a conglomeration of professions in order to produce sustainable cities of the highest standard. With the current ideology of architecture, architects cannot work with other professionals like agronomists, engineers and environmentalists in coming up with a design as the Architects Act (Chapter 27:01) prohibits non-registered persons from working with registered architects.

One of the important things required for an effective role of the built environment disciplines and practice in the implementation of the NUA is Building Information Modelling (BIM). BIM is a process of creating a model that contains all the required information to deliver a project. The most important aspect about BIM is the desire to create a Common Data Environment (CDE), a shared repository to which relevant stakeholders have access (Succar 2009). BIM allows continuous addition of information from project onset to closure and beyond. It also allows the client to have a complete model with all the information about the project, allowing its use for maintenance, marketing, and any other purpose that requires building information (Borrmann et al. 2018).

BIM provides a tool for the design of a city not to be confined to an individual but a multidisciplinary team in order to minimise errors and omissions. It allows stakeholders to work concurrently on a project and distributes responsibilities fairly. Each stakeholder is able to input information in the central database and its effect will be visually realised and appropriate decisions can be taken before construction takes place. The role of architects, engineers and construction consultancies has to change from a system of distinct specialised professionals to collaborative interactive field of professionals. For the NUA to bring positive results, BIM implementation in Zimbabwe should be encouraged as opposed to the current setting of fragmented discrete working.

Model Building By-Laws of 1977 needed to be regularly updated in sync with changing building technology so as to minimise the need for developers to obtain approval from the Standards Association of Zimbabwe each time they want to use new materials in construction. Traditional building materials were dominated by concrete, steel and timber. A paradigm shift is required in order to allow contemporary thinking, methods and practices in building technology. In this day and age where structural glass is as strong as steel (Wittenberg and Krynski 2010) and structural wood as strong and fire resistant as concrete, one cannot imagine a whole plan failing to be approved by the local planning authority because it used structural glass or wood where traditionally it was impossible.

The role of architecture in implementing the NUA cannot be overemphasised; however, architects need to be better equipped for the task. In the area of architecture education, there is a need to expose students to the latest technology and global best practices in architectural design. Building climate resilience in architectural design and in the use of construction materials has drawn attention to locally available resources. The use of alternative technology is expected to improve access to housing. Training opportunities are critical to participation of local communities and all levels in the building industry. All these endeavours would need to be supported by macroeconomic stability that renders the built environment a viable vehicle for investment.

This chapter has shown that the architectural profession in Zimbabwe has not provided adequate scope for collaboration with other built environment professionals thereby sub-optimising architecture's contribution to the achievement of the Sustainable Development Goals. However, it is clear that architecture has an important role in the implementation of the SDGs. The research recommends an interdisciplinary approach in the built environment whereby the creation of new cities, neighbourhoods and housing areas should not be left to discrete professions but be done through collaborative and interdisciplinary efforts.

References

Ali M (2021) Urbanisation and energy consumption in Sub-Saharan Africa. Electr J 34(10):107045
Angelopulo G (2021) A comparative measure of inclusive urbanisation in the cities of Africa. World Dev Perspect 22:100–313
Barkman RC (2021) The importance of outliers: clues to unforeseen patterns. Psychol Today
Borrmann A, Konig M, Koch C, Beetz J (2018) Building information modeling: why? What? How? In: Borrmann A, Konig M, Koch C, Beetz J (eds) Building information modeling: technology foundations and industry practice. Springer, pp 1–24
Brussat D (2020) Lesson plan # 9: ideas on architectural education. In: Richards K, Salingaros N (eds) On the future of architectural education. https://patterns.architexturez.net/doc/az-cf-194150 Accessed 21 Aug 2023
Caprotti F, Cowley R, Datta A, Broto VC, Gao E, Georgeson L, Herrick C, Odendaal N, Joss S (2017) The new urban agenda: key opportunities and challenges for policy and practice. Urban Res Pract 10(3):367–378
Chigudu A (2021) The changing institutional and legislative planning framework of Zambia and Zimbabwe: nuances for urban development. Land Use Policy 100:104–941
Chombo IM, Youm SH (2021) Building performance evaluation of public housing in Harare, Zimbabwe. Acta Structilia 28(2):1–22
COR (2022) What are the roles and responsibilities of an architect in construction. https://projectcor.com/what-are-the-roles-and-responsibilities-of-an-architect-in-construction/. Accessed 27 Feb 2023
De Waal A, Weaver M, Day T, van der Heyden B (2019) Silo-busting: overcoming the greatest threat to organisational performance. Sustainability 11(23):6860–6891
Desmond M, Gershenson C (2016) Housing and employment insecurity among the working poor. Soc Probl 63(1):46–67
Dolman N (2021) Integration of water management and urban design for climate resilient cities. In: de Graaf-Van Dinther R (ed) Climate resilient urban areas. Palgrave Macmillan, Cham, pp 21–43
Dovey K, Van Oostrum M, Chatterjee I, Shafique T (2020) Towards a morphogenesis of informal settlements. Habitat International 104:102240, https://www.sciencedirect.com/science/article/pii/S0197397520303714. Accessed 20 Aug 2023
Freilich M (1967) Ecology and culture: environmental determinism and the ecological approach in anthropology. Anthropol Q 40(1):26–43
Gold JR (2013) The experience of modernism: modern architects and the future city, 1928–53. Routledge: Taylor and Francis Group
Govender T, Barnes J, Pieper C (2011) Housing conditions, sanitation status and associated health risks in selected subsidised low-cost housing settlements in Cape Town, South Africa. Habitat Int 32(2):335–342
Granberg M, Glover L (2021) The climate just city. Sustainability 13(3):1201
Hegemann W, Peets E, Myers T (2008) The American Vitruvius: an architects' handbook of civic art. De Facto Publishing, New York
Highet J (2013) African Architecture: 21st century "Afritecture", New African. https://newafricanmagazine.com/4145/. Accessed 19 Aug 2023

Hind C, Orsini F, Pugh S (2022) The architecture drawing book: RIBA collections. RIBA Publishing, London
Horne J (2011) Architects, stadia and sport spectacles: notes on the role of architects in the building of sport stadia and making of world-class cities. Int Rev Sociol Sport 46(2):205–227
Jarman D, Oborn P, Walters JG (2018) Planning for rapid urbanisation: survey of the architectural profession in the commonwealth, London, UK
Koch M, Worndl W (2001) Community support and identity management, conference: proceedings of the seventh European conference on computer supported cooperative work, 16–20 Sept 2001, Bonn, Germany
Matamanda AR, Nel V, Chirisa I (2021) Introduction to the Urban Geography Scape of Zimbabwe. In: Matamanda AR, Nel V, Chirisa I (eds) Urban geography in postcolonial Zimbabwe: paradigms and perspectives for sustainable urban planning and governance, Springer, Cham. Switzerland. pp 1–15
Mishra PC (1998) Right to Shelter: a human right perspective. J Indian Law Inst 40(1–4):230–242
Mtomba V (2015) Multi-currency revives housing project (The Standard, 5 April 2015). https://allafrica.com/stories/201504050341.html. Accessed 19 Aug 2023
Neuman M (2021) The emergence of regional design: recovering a great landscape architecture and urban planning tradition. In: Neuman M, Zonneveld W (eds) The Routledge handbook of regional design. Routledge, New York, pp 1–14
Perez-Gomez A (1982) Architecture as drawing. J Architectural Educ 36(2):2–7
Richards PJ (1979) Housing and employment. Int Labour Rev 118(1):13–20
Rudofsky B (1964) Architecture without architects, an introduction to nonpedigreed architecture. The Museum of Modern Art, Garden City, New York
Sadler RC, Bilal U, Furr-Holden CD (2021) Linking historical discriminatory housing patterns to the contemporary food environment in Baltimore. Spat Spatio-Temporal Epidemiol 36(4):100387
Saunders MN, Lewis P, Thornhill A (2019) Research methods for business students. Pearson Education Limited, Harlow UK
South African Institute of Architects (2020) A client's guide to working with a SAIA architect. South African Institute of Architects, Johannesburg, South Africa
Succar B (2009) Building information modelling framework: a research and delivery foundation for industry stakeholders. Autom Constr 18(3):357–375
Tafuri M (2006) Interpreting the Renaissance: princes, cities, architects. Yale University Press, New Haven: CT
Tibaijuka AK (2005) Report of the fact-finding mission to Zimbabwe to assess the scope and impact of Operation Murambatsvina by the UN special envoy to Zimbabwe, Nairobi, UN-Habitat. http://hrlibrary.umn.edu/research/ZIM%20UN%20Special%20Env%20Report.pdf. Accessed 19 Aug 2023
Ulusoy M, Kuyrukcu EY (2012) The meaning and importance of the traditional architecture in architecture education; Gonen Winter School Model. Procedia-Soc Behav Sci 47:2120–2126
UN-Habitat (United Nations Human Settlements Programme) (2014) The state of African cities 2014: re-imagining sustainable urban transitions, United Nations Human Settlements Programme, Nairobi, Kenya
United Nations (2017) New urban agenda, Habitat III, Quito 17–20 October Qui 2016, United Nations conference on housing and sustainable urban development
Wittenberg D, Krynski M (2010) Design of a single span nine metre long glass bridge. In: Bos, Louter, Veer (eds) Challenging Glass 2—conference on architectural and structural applications of glass, TU Delft, Netherlands
Zajda J (2021) Third international handbook of globalisation, education and policy research

Sustainable Urban Mobility and the New Urban Agenda in Zimbabwe

Tinashe Kanonhuhwa, Nesbert Mashingaidze, and Herbert Chirwa

Abstract Continued urbanisation has become a common phenomenon in most urban setups and Zimbabwe is no exception. Apart from the unplanned mushrooming of informal urban settlements, the Zimbabwean government has also been faced with the need to formally accommodate the masses through the provision of affordable housing. The formation of new settlements has, however, not tallied with continuous improvements in service delivery, particularly transport and associated infrastructure, and this has culminated in a series of problems such as in increased pressure on urban roads, increased journey times and poor accessibility. Through an archival method of data collection and analysis depending highly on outcomes of doctoral research, this book chapter analyses how the Zimbabwean government has moved towards the attainment of the Sustainable Development Goals and the New Urban Agenda in relation to the improvement of the urban transport system and infrastructure to allow for smooth circulation of traffic, particularly public transport. The book chapter reveals challenges caused by an inefficient public transport system such as congestion, increased inaccessibility of land uses and travelling delays suffered by road users as they try to reach various destinations. This chapter positively contributes to the field of Town Planning by unpacking the role of the New Urban Agenda in positively influencing progressive developments in Zimbabwe's public transport system.

Keywords Built environment · Mobility · New urban agenda · Public transport system · Rapid urbanisation · Sustainable development

T. Kanonhuhwa (✉) · N. Mashingaidze · H. Chirwa
Department of Demography Settlement and Development, University of Zimbabwe, Box MP 167, Mt. Pleasant, Harare, Zimbabwe
e-mail: tnkanonhuhwa@gmail.com

H. Chirwa
Department of Architecture and Real Estate, University of Zimbabwe, Harare, Zimbabwe

C. Chavunduka and I. Chirisa (eds.), *New Urban Agenda in Zimbabwe*,
Advances in 21st Century Human Settlements,
https://doi.org/10.1007/978-981-97-3199-2_8

1 Introduction

Increased urbanisation has become a common global phenomenon. This is evidenced by the growth in urban population which shows a consistent increase during 1950–2050 (Liu et al. 2018; Kundu and Pandey 2020). Rapid urbanisation in sub-Saharan Africa, of which Zimbabwe is an important part, is a reality and cannot be wished away. It has been observed that rates of urbanisation in Africa are the highest in the world (Leeson 2018; Kundu and Pandey 2020). And by 2025, more than half of the African population will be living in urban areas. In this regard, some scholars have argued that a shrinkage in service provision has been experienced as urban population grows, and this has been coupled with the observation that the formal transport system that dominated most African landscapes has disappeared as the number of people increases in the urban sphere (Mbara and Pisa 2019). Continued urbanisation has been linked to various transport-related challenges in the form of increased transport demand, hiking of fares, creeping in of unregistered transport operators commonly known as *mushika-shika* (informal public transportation system in Zimbabwe) and shrinkage in formal transport as well as dilapidation of transport infrastructure among other challenges (Peeters et al. 2021; Baindur and Rao 2016; African Development Bank 2019). In the process, people have in turn endured long waits and increased walking distance to areas of interest due to inadequate transport (Kumar et al. 2021). Traffic congestion has increased due to poor road infrastructure and consequently leads to direct and indirect environmental degradation in the form of increased emissions of greenhouse gases and unauthorised extension of road shoulders in trying to avoid congested roads. It has been observed that there is a mismatch between the increased development of urban settlements and the upgrading of the road system to meet transport demand.

Following the rapid development of settlements (planned or unplanned), the increase in formal public transport has not tallied with an increase in travel demand on Zimbabwe's urban roads. This has resulted in increased pressure on urban roads and its infrastructure, compounded by increased mushrooming of informal transport operators to harness the increase in demand (Mbara and Pisa 2019). Apart from the positive role played by the informal sector in partly addressing the increase in demand for transport, the informal transport system has been linked to unsafe road practices such as reckless driving, rogue behaviour by informal minibus drivers that are believed to have increased fatalities on urban roads (Zeeman and Booysen 2013).

Most of the informal public transport (IPT) is not regulated and is associated with reckless driving on the part of the informal transport operators, and due to the sour relations between law enforcement agencies and informal transport operators, some of the informal taxis on the roads in the city of Harare have been seen moving around with smashed windscreens (Tichagwa 2016; Middleton 2018). This in turn makes the city an unsafe place to be when it comes to effective transport circulation and the safe transportation of the public. Also, due to little or no improvement in the road infrastructure to match the demand, most of Zimbabwe's urban roads, particularly in Harare, have become littered with potholes (Mbara and Pisa 2019). The state of the

Zimbabwean road system does not in turn tally with the Sustainable Development Goals (SDGs), particularly Goal 11 which aims at creating sustainable cities and communities.

The New Urban Agenda (NUA) was adopted on October 20, 2016, and with this, cities would be expected to follow laid down standards and principles as regards to development, management, planning and improvement of urban areas (UN-Habitat 2020). The NUA would in turn be viewed as an accelerator of SDG 11 which aims to make cities inclusive, safe, resilient, and sustainable (UN-Habitat 2020). It would touch on many aspects on the Zimbabwean road system from transport operators, and the safety of the modes themselves up to the need to make the road infrastructure safe for use and in good condition. These conditions have a bearing on sustainable urban mobility. This calls for the Zimbabwean government through its ministries and departments to intensify the development and improvement of roads and related infrastructure in line with the NUA. This will go a long way in achieving sustainable urban transport and consequently sustainable urbanisation.

2 Background and Overview

The challenges in Zimbabwe's transport system can be traced back to 1980 when the country attained its independence. To reduce social inequality, in 1988 the Zimbabwean government attained a 51% stake in the Zimbabwe United Passenger Company (ZUPCO), and after implementation of the Economic Structural Adjustment Programme in 1991, public transport was deregulated in 1993 (Mbara 2006; Nyatondo 2013; Dumba 2017). This paved the way for informal public transport operators in the urban landscape. The operators' licences for commuter omnibuses became easier to get and this allowed small players to come and fill the demand gap left by the conventional public transport system (Nyatondo 2013). Apart from the informal transport system, people have also found a cheaper market for cars from Japan, and this has resulted in increased importation of cars in urban areas, especially Harare (Bandauko et al. 2016). This has in turn increased pressure on urban roads and hence, the resultant increase in potholes as witnessed on most of Harare's roads (Zikhali 2017; Mbara and Pisa 2019).

Despite all these challenges, Zimbabwe's road-based transport system remains a mode of choice regionally due to the key role it plays in the Southern African Development Community (SADC) rail and road corridor, as evidenced by its 87% contribution to the visitor's arrivals in 2016/17, compared to the 13% arrivals by air in Zimbabwe within the same year (African Development Bank 2019), indicating that Zimbabwe's road system is not only essential to local people, but regionally as well. Today, Harare's transport woes range from transport shortages to poor road infrastructure (African Development Bank 2019). The decline in the road network has been attributed to the lack of funding for periodic and routine maintenance in the 1990s up to the present time (Bandauko et al. 2016). The book chapter aims to examine the role of the New Urban Agenda in facilitating transportation, mobility, and accessibility with reference to road improvements in Harare. Priority shall now be given to the objectives of the study.

The objectives of the study are to:

(a) Examine the state of Zimbabwe's urban transport system in relation to the New Urban Agenda.
(b) Explore some of the transportation, mobility and accessibility challenges being experienced in Harare's metropolitan area.
(c) Evaluate how the NUA can be used to solve these challenges and improve Zimbabwe's urban roads.

3 Methodology

The study adopted the nested approach suggested by Sexton (2000). The researcher used interpretivism as a research philosophy. This has enabled the researcher to explore and interpret the contribution of the New Urban Agenda to sustainable mobility and accessibility as well as provide philosophical underpinnings for qualitative studies which underpin this study. Exploratory sequential method was used to explore qualitative data on issues of sustainable mobility, accessibility in relation to the New Urban Agenda. The study utilised two research designs, namely case study and documentary review paying particular attention to allied doctoral research that elaborates issues under discussion. Case study design enabled the researcher to explore issues surrounding sustainable mobility and accessibility with reference to Harare whilst documentary review enabled the researcher to review relevant literature on sustainable mobility, circulation and accessibility in relation to New Urban Agenda. The researchers collected both primary and secondary data. Primary data was collected using qualitative research methods. These included key informant interviews using an interview guide and observations. Purposive sampling was used for the study so that a specific group of individuals were selected to ensure better insights and more precise research results (Mweshi and Sakyi 2020; Klar and Leeper 2019). In addition, key informants have vast experience and in-depth understanding of mobility, and infrastructure in urban areas. Key informants were drawn from the City of Harare transport division, Government officials such as those from the Ministry of Transport and Infrastructure Development, heads of parastatals such as Zimbabwe National Roads Authority (ZINARA), Zimbabwe Energy Regulatory Authority (ZERA) and Environmental Management Agency (EMA). The book chapter comes closer home by singling out the case of Harare to understand transport and mobility in urban regions. Harare has been chosen because it is the capital city and has experienced rapid urbanisation and more importation of cars than other cities in the country (Bandauko et al. 2016). The research unpacks transport mobility in Zimbabwe's urban areas as well as the challenges encountered. It also views the issues in the 'lens' of the NUA by assessing what can be done to improve traffic circulation and mobility in Zimbabwe's urban landscape. The book chapter also explores possible local solutions to improve traffic-related challenges under the New Urban Agenda to facilitate the sustainable development of cities.

4 Study Area

Motivated by the need to find a better livelihood in the city, the city of Harare has experienced a significant rural–urban migration (Marondedze and Schütt 2019) and hence the increase in transport demand. Harare was selected as the study area due to the city facing continuous mobility and urban challenges. These include increased mushrooming of informal public transport on the urban landscape, ever-increasing population with associated increased transport demand, increased fights between law enforcement agencies and informal public transport operators (thereby putting the lives of commuters at risk) (Bandauko et al. 2016; Mbara and Dumba 2017; Mbara and Pisa 2019; Mupfumira and Wirjodirdjo 2015; Tichagwa 2016). In addition, poor air quality has also been experienced due to the release of exhaust fumes from old vehicles at illegal ranks, and the informal public transport has also been associated with the charging of higher fares, as well as the unsafe picking and dropping of people at undesignated places (Mbara and Dumba 2017), thereby requiring stricter government controls if a safer transport environment is to be created.

Harare is the capital of Zimbabwe. It is situated at an elevation of 1494 m above sea level and is located at − 17.83 latitude and 31.05 longitude. Harare had 2,427,231 residents as of the 2022 census, which is 12.2% of the country's total population and represents an average growth rate of 2.1% (Zimstat 2022). Map 1 shows the spatial form of Harare.

5 Conceptual Framework

The study adopted the A-S-I approach as a model towards sustainable transport, mobility and accessibility in Zimbabwe. The approach integrates important aspects of urban transport as well as meeting the traits and tenets of the New Urban Agenda as it integrates the economy, society and mobility. As a people-centred sustainability approach, the Avoid/reduce-Shift/maintain-Improve (A-S-I) concept has gained momentum in the New Urban Agenda. The avoid or reduce refers to the need to improve the efficiency of the transport system. It is argued that through transport oriented and compact development of cities, the need for motorised travel and then trip length can be reduced. In the planning and designing of settlements, residential, work and leisure districts must become more closely connected and intermixed to address transportation, mobility and accessibility challenges.

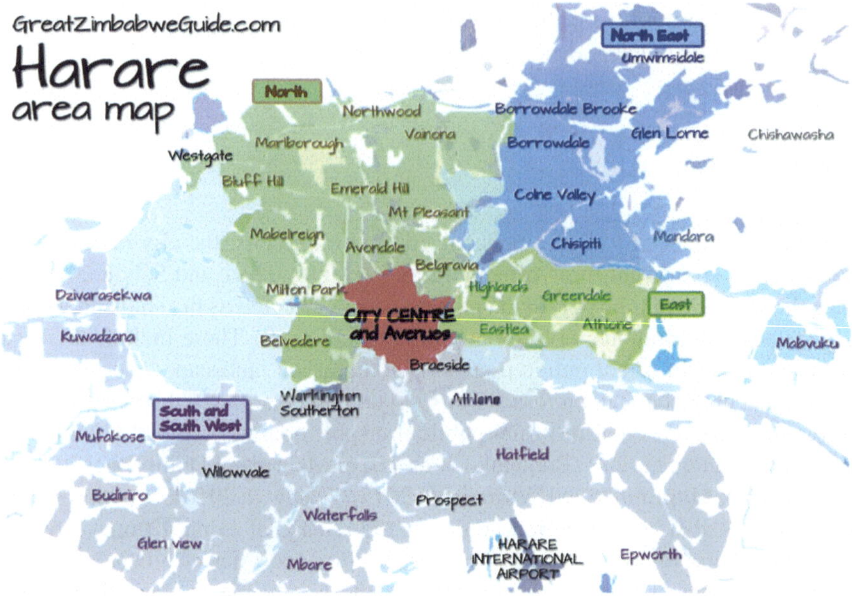

Map 1 The map for the city of Harare. *Source* https://www.greatzimbabweguide.com/areas-of-harare/

The shift or maintain instruments seek to improve individual trip efficiency. A shift from the most energy consuming and polluting urban transport mode, that is cars, towards more environmentally friendly modes addresses the motorised transport system challenges. Lastly, the improve pillar focuses on vehicles and fuel efficiency as well as on the optimisation of the operational efficiency of public transport. This includes the attractiveness of public transport. Since the New Urban Agenda focuses much of its attention on sustainability issues and makes human settlements sustainable, improvement of the energy source required for the smooth operation of transport is crucial. However, for energy efficiency to be achieved, a holistic approach to urban transport and mobility management is important (UN-Habitat 2016). In this study, the framework is used for purposes of understanding the link between the urban transportation question and the New Urban Agenda in Zimbabwe. It seeks to achieve significant GHG emissions reduction and create more livable cities.

6 Literature Review

6.1 Mobility and Accessibility Challenges in Cities: A Global Perspective

Present-day cities have been characterised by lots of challenges with social, economic, political, and environmental dimensions. Some of these challenges have been exacerbated by urbanisation. These challenges include both pedestrian and vehicular traffic congestion resulting in the inaccessibility of some parts of the city. Coupled with this are the various forms of pollution such as land, air, water, noise and visual pollution which are both urban doldrums. Gkoumas et al. (2021) add that cities today are challenged with meeting steadily increasing mobility demands and providing additional road space has not helped to solve the pressing problems of unacceptable levels of congestion, environmental air quality deterioration and increasing Green House Gas (GHG) emissions. Because of all these urban nuisances, cities have not become more liveable and mobility in and around cities remains characterised by increased journeys to work making cities not human-centred as some activities are becoming unsustainable and do not meet the demands of users. However, the quest for improving urban mobility through sustainable transport has remained one of the top priorities of New Urban Agenda (NUA).

However, Turok (2015) posits that nations are still far from adequately addressing these and other existing and energy challenges. To achieve sustainable mobility, transportation, and accessibility in urban areas, planning and designing of mobility and transport infrastructure is crucial. This entails incorporating elements of sustainability in the planning and designing of transport infrastructure. Therefore, the New Urban Agenda should not be just an abstract principle but a principle that needs to be cultivated from the planning and designing stages. Studies revealed that urban problems such as inequalities, poverty, and hunger can be addressed by the way cities and human settlements are planned, designed, and financed (Turok 2015). It is through proper urban planning and designing of transport and mobility infrastructure that sustainability, inclusivity, and environmental protection can be achieved. Moreover, the United Nations Human Settlements Programme (UN-Habitat, 2020; United Nations 2017) and United Nations Centre for Regional Development (UNCRD) are in support of the full provision of safe, age and gender responsive, affordable, accessible and enabling meaningful participation in social and economic activities in cities and human settlements, by integrating transport and mobility planning into overall urban and territorial plans and promoting a wide range of transport and mobility options.

6.2 New Urban Agenda and Mobility

According to UN-Habitat (2017), the New Urban Agenda presents a paradigm shift based on the science of cities. It is also viewed from the perspective of systems in urban areas. This means that the New Urban Agenda covers a range of issues and sustainability remains the key issue. It is asserted that the New Urban Agenda lays out standards and principles for planning, construction, development, management, and improvement of urban areas along with its five main pillars of implementation, that is, national and urban policies, urban legislation and regulations, urban planning and design, local economy and municipal finance and local implementation (UN-Habitat 2017). The developmental challenges which have been presented by urbanisation today tempted local authorities and global leaders to think of the New Urban Agenda that aims to address urban problems and plan make cities liveable and achieve human settlement goals. The New Urban Agenda incorporates a new recognition of the correlation between good urbanisation and development so that negativity associated with urbanisation will be outweighed by the benefits which include good living. However, it is important to note that the New Urban Agenda is an action-oriented document which sets global standards of achieving sustainable urban futures, rethinking the way we build, manage, and live in cities through cooperation with committed partners, relevant stakeholders and urban actors at all levels of government as well as the private sector. In relation to mobility, transportation and accessibility, the transformative commitments of the New Urban Agenda include the commitment to sustainable and efficient transport for achieving the benefits of connectivity and reducing the financial, environmental, and public health costs of congestion and air pollution (Palop 2020). However, it is argued that the uptake of electric mobility in all forms, in the context of better and compact urban planning has the potential to make an important contribution to a low-carbon development pathway for transport (Palop 2020).

6.3 Environment, New Urban Agenda and Mobility

Additionally, the NUA is a blueprint for UN-Habitat and its partners in government. Several stakeholders ranging from the top supranational bodies such as the United Nations, civil society, communities, the private sector, professionals, the scientific and academic community are in support of the 2030 agenda on sustainable development, with main reference to Sustainable Development Goals number 11 that aims at making cities inclusive, safe, resilient, and sustainable (Turok 2015). The NUA is anchored on the following principles: Governance structures that aim to have participatory urban policies that mainstream sustainable urban and territorial development as part of integrated development strategies and plans. Social inclusion is also among the principles of NUA. It entails that development must protect the planet and enable all inhabitants, whether living in formal or informal settlements to lead decent, dignified and rewarding lives and to achieve their full human potential. Spatial development is also one of the main principles of NUA. It advocates for balanced territorial development that accounts for different scales of cities and human settlements and strengthens their role in food security and housing infrastructure. Urban prosperity and environmental sustainability are among the important principles of NUA, relating to urban climate change discourse, urbanisation through its industrial emission of GHG and fumes from vehicle exhaust (Mashingaidze et al. 2021). This contributes to global warming (worldwide increase in global temperatures) threatening the quest for sustainable urban environments, transportation and mobility.

Through environmental sustainability, the NUA is encouraging the reduction of unsustainable consumption and production patterns, discouraging loss of biodiversity, pressure on natural resources since these natural resources are the mother nature and form the basis for human settlements. It also deals with man-made disasters and climate change and its related risks, undermining efforts to end poverty and to achieve sustainable development. Surrounding the whole discussion is the issue of sustainable development that is more pronounced by the NUA.

However, Turok (2015) states that the persistence of multiple forms of poverty, growing inequalities both local, national and regional and environmental degradation remain among the major impediments to the quest for sustainable development worldwide. Moreover, social and economic exclusion and spatial segregation often exert an irrefutable pressure on the reality of cities and human settlements. Transport infrastructure which occurs in various forms such as roads, rail and air remains the bedrock in city mobility and circulation. Transport infrastructure is a key enabler of mobility in cities. However, with the NUA, transport sustainability is on the cards. To address sustainability concerns in urban areas, transport sustainability remains the key enabler. This is because of environmental disturbances in urban areas that tempted local authorities to think of environmental sustainability as an important component. The NUA is multi-sectoral and trans-discipline in nature; therefore, the mobility and transport sector cannot be left out. However, to address some of the sustainability challenges related to mobility and accessibility in urban areas, the European Commission is working on a new European Union urban mobility framework to support member states, regions, and cities as they develop safe, accessible, inclusive, smart, resilient, and zero emissions urban mobility (Gkoumas et al. 2021). It is added that all these measures contribute to the European Union climate and transport policy objectives and targets are set out in the European Green Deal and sustainable and SMART Mobility strategy (Gkoumas et al. 2021).

The quest for achieving sustainable transport infrastructure in urban areas remains rhetoric and with many puzzling explanations. Behruz et al. (2013) state that there are a lot of explanations for sustainable development but the nature or degree of them can be the same or different depending on the use of sustainable resources. It is further argued that in developing countries, in particular, cities have experienced a rapid growth in transport-related challenges and problems (Cervero 2013). These problems range from economic to social problems. These include pollution, congestion, accidents, public transport decline, environmental degradation, climate change, energy depletion, visual intrusion, transport disorders and lack of accessibility for the urban poor. In developed countries, particularly in Northern Europe, some cities have witnessed a trend of reclaiming urban space from the automobile and prohibiting cars from major parts of downtown areas and or confining them in other ways. These have been seen as sustainable ways of containing unsustainable urban transport in developed countries. To date, the places are often considered as leading examples of sustainable urban development, as cities across the world strive to meet urban sustainability standards by improving public transport and encouraging non-motorised modes. Other methods such as creating pedestrian zones limiting the use of private cars have been deployed to control traffic. It can be noted that, in Northern Europe city transformations have been associated with automobile dominance. However, these require the integration of Intelligent Transportation Systems (ITS) in dealing with urban transport and for urban transport to be suitable.

As alluded above, the New Urban Agenda envisaged that cities and human settlements must fulfil their social function, including the social and ecological function of land, with a view to progressively achieve the full realisation of the right to adequate housing as a comment of the right to an adequate standard of living, universal access to safe and affordable drinking water and sanitation as well as equal access for all to public goods and quality services. Mobility and transportation are crucial in the development of urban areas as this enables the function of an urban settlement. However, the challenge in domesticating the urban agenda is the state of preparedness of our cities. The need to address mobility and transportation challenges in urban areas is crucial. It is suggested that cities are challenged with meeting steadily increasing mobility demands. It is argued that cities develop policies that are grounded in the context of local realities and do not get seduced into imitating satellite cities, flagship real estate projects and other grandiose urban models pursued by wealthier nations which are now giving rise to a myriad of urban challenges in relation to transport and mobility (Watson et al. 2019). Relating to urban mobility, circulation and transportation, urban mobility and metropolitan; transportation systems are at the core of the most significant challenges faced by our cities in the first part of the twenty-first century (McCann 2015).

6.4 Policy Development and Mobility

Behruz et al. (2013) proffer traits and tenets of sustainable transport systems which include the needs of individuals, companies and society to be met safely and in a manner consistent with human and environmental, health and provide equity within and between successive generations. Economically, a sustainable transport is affordable, operates fairly and efficiently, offers a choice of transport mode and supports a competitive economy and overall, a balanced regional development. Additionally, sustainable transport limits emissions and waste within the planet's ability to absorb them and uses renewable resources at or below the rates of development of renewable substitutes, while minimising the impact on the use of land and the generation of noise. However, the corresponding increase in greenhouse gas emissions that presents a clear challenge to the international community requires policies that promote innovation technologies at local, national and regional scales. Mobility policy is a part of a complex pattern which needs coordinated vision, commitment and investments to have visible results, safer and reliable transport networks as well as safer and performing vehicles.

However, several raft measures have been put in place to address these challenges. Some of these measures which have been proposed include alternative fuels and infrastructure development. It is suggested that the market uptake of alternative fuel vehicles and infrastructure is expected to accelerate in the coming years, driven by the need for member states and automotive manufacturers to comply with relevant legislative provisions for reducing GHG emissions from transport, including the recently revised carbon dioxide emission performance standards for cars and vans and heavy-duty use vehicles (Gryparis et al. 2020). It is further added that other modes of transport will have to increasingly deploy alternative fuels and electric power trains and related infrastructure such as inland waterways, maritime, rail or aviation transport.

At the national level, the Government of Zimbabwe has initiated policies and strategies aimed at promoting sustainable transport systems in its urban areas and these include the National Transport Policy (NTP) of 2012 and the National Development Strategy 1 (NDS 1) of 2020. The NTP aimed at promoting a reliable and cost effective, safe and sustainable road system with one of its overall objectives being inclined towards minimising the loss of life and property, as well as damage to the environment, and also towards attaining social needs through securing a reliable level of mobility for the population. In the same spirit, the NDS 1 of 2020 was an instrument that aimed to achieve Zimbabwe's vision 2030, which aimed at achieving an empowered and prosperous upper-middle society by the year 2030. NDS 1 is a five-year plan that covers the years 2021–2025, and it acknowledges that the poor condition of Zimbabwean roads indirectly and directly impacted road transport safety, and this has resulted in increased road accidents. NDS 1 aims at achieving a high quality and efficient public transport service in both rural and urban areas, and this shall be achieved through insertion of road signs on new and existing roads, installation of road signs that are more visible and less prone to vandalism, as well as implementation of traffic management signs that are in line with regional standards and also ensure that possible engineering deficiencies that may be on the road 'black spots' are rectified (Government of Zimbabwe 2020).

7 Results and Discussion

7.1 Public Transport System in Harare

The City of Harare Strategic Plan 2012–2025 has the objective of creating an integrated living environment that satisfies the 'live, work and leisure' concept. A sustainable city will have effective and efficient mobility systems implemented through an energy-efficient mobility master plan. But the city is yet to develop an energy-efficient mobility master plan. However, the rationale of the New Urban Agenda in urban transportation, mobility and accessibility is to promote sustainable urban mobility in different contexts and identifies infrastructure and policy measures with the highest mitigation potential. These policies and measures will help to reduce the emission of GHGs and strive for sustainable human settlements.

The study findings reveal that the city of Harare is facing a myriad of transport challenges ranging from practical to policy ones (Dumba 2017). Such challenges limit the city in quenching its thirst towards achieving sustainable transport mobility and accessibility (Mudawu 2021). The general state of public transport in the city of Harare is characterised by chaos as there are many uncoordinated public transport operators, namely the Zimbabwe United Passenger Company (ZUPCO), private commuter omnibus, private vehicles mainly operating as mushika-shika and registered taxis (Dumba 2017). However, Dumba (2017) notes that operators of mushika-shika do not operate from designated points resulting in some streets becoming congested, and there is reduced access to some points. Traffic congestion is a worldwide problem that is restricting access in cities, and have major impacts on quality of life.

Exacerbating the situation, the state of some vehicles which are used to ferry people is not road-worthy, thereby compromising the quality of service provided by public transport operators (Chikengezha and Thebe 2022). In trying to show the increased number of public transportation vehicles within the city of Harare, Dumba (2017) compares the capacity of the different ranks within Harare's CBD against their actual capacity as follows: the design capacity for Charge Office rank was set at 550 minibuses an hour; however, it accommodates 2884 minibuses an hour. That of Fourth Street rank was set at 580 while it now accommodates 2413 per hour. The rank at Market Square was set at 240 minibuses an hour, however; it accommodates 3533 vehicles an hour. It has been observed that some existing transport infrastructure such as Copacabana, Fourth Street, Charge Office and Market Square bus ranks can no longer accommodate more combis and buses as they are currently used by ZUPCO commuters and buses. This has resulted in some commuters resorting to pick and ride service along the main streets of Harare, leading to a chaotic situation in the city.

The study observed that some roads in the Central Business District (CBD) have poor signage and are not well-marked compounding the problems of accidents and congestion. Other causes of congestion include poor traffic control systems, poor driver behaviour, dilapidated and narrow road infrastructure, shortage of public transport (buses) and unlicensed pirate taxis (mushika-shika) (Matenga 2021; Zimeye

2021; Mudawu 2021). This compromises the ability of the city to achieve sustainable transport mobility and accessibility. Coupled with this, it has been observed that hard infrastructure failure and weak enforcement of traffic laws by joint operation of the Zimbabwe Republic Police (ZRP) and municipal police had resulted in some accidents, especially at intersections controlled by traffic lights.

7.2 *Urban Transportation, Mobility, and Accessibility Challenges in Harare*

7.2.1 Urban Form

Some of the mobility and accessibility challenges in Harare have been ascribed to urban form. Harare is a monocentric city with arterial roads converging on the CBD. Land use zoning laid out the city in discrete commercial, industrial, institutional, residential and recreational zones which generates demand for travel from one land use zone to the other. Rapid urbanisation in Harare got uneven as it led to rapid residential expansion as compared to other facets of urban development. City expansion has gone beyond the municipal boundary as many residents sought residential plots in the urban periphery where state land is affordable. Development in the city periphery has been low density, single-use and fragmented and in the absence of a reliable public transport system, commuters are mostly car-dependent (Litman and Steele 2017; Dumba et al. 2017).

A key informant from the Zimbabwe National Roads Authority (ZINARA) bemoaned the automobile dependency which has been created by the separate spatial location of land uses, particularly the distance between residential and workplaces which eventually determined the frequency and mode of travel in the city. The city form has pushed many households to buy cars even though they can barely afford them. It is a calculated move taken by commuters to solve mobility challenges in the face of unreliable and mostly unavailable public transport system.

7.2.2 Public Transport System Accessibility

Transport system accessibility in Harare was found to be sub-optimal. Transport system accessibility is the ability of users to reach key destinations in a single trip (Batty and Marshall 2017). It also refers to the availability and usability of the transport modes available for use by commuters. A city of Harare transport planner believed that accessibility of the transport system in Harare is not favourable. His main concern was the lack of integration of the different types of transport. He was of the view that an interchange system would afford people enough connectivity in the Harare metropolitan area.

Since the onset of the coronavirus pandemic in 2019 (COVID-19), the government has restricted all urban transportation in the country to be done by Zimbabwe's regulated public transport system known as ZUPCO registered buses. This has resulted in long queues at stations, as buses have not been able to cope with transport demand and this has also been the case for other cities such as Bulawayo (The Chronicle 2020). However, the government initiative of controlling both long- and short-distance public transport through ZUPCO is noble but this is coming at a time when the government is battling with lots of challenges such as the COVID-19 pandemic and the ill-performing economy. A member of the Urban Transport Association had this to say, 'The ZUPCO buses are insufficient, of poor condition and have no fixed timetables just like our members. In some cases, surprisingly, the buses do not come, leaving commuters stranded. This has increased transport unreliability'. If there were employment and business opportunities in close proximity to residential areas, there would be no need for them to travel to the city centre. A policy that creates employment opportunities in residential areas can reduce the volume of traffic into the city centre or can promote modal shift from motorised to non-motorised modes of transport. The Public Private Partnership (PPP) agreement that is prevailing in the transport sector is viewed as a good initiative towards sustainable urban transport as more buses and Kombis from the private sector are complimenting governments' initiative in providing public transport. Recently, the government of Zimbabwe has taken a further step in improving the movement of people, especially during peak periods by enabling National Railways of Zimbabwe (NRZ) to operate urban transit under the auspices of ZUPCO.

7.2.3 Public Transport System Inclusivity

Field observation confirmed the fact that the public transport system in Harare is not inclusive. A representative of the Ministry of Transport and Infrastructure Development indicated that urban transport facilities have not been tailored to cater for the mobility needs of people living with disabilities. Very few of the buses and minibuses in operation have facilities for people living with disabilities, especially wheelchair users. In low-income areas, there are limited or no cycle tracks and pedestrian walkways, let alone wheel tracks for wheelchair users in a planning system that has catered for motorised transport.

Transport costs have been an inhibiting factor for poor individuals to access opportunities and services that uplift their lives. Transport poverty has been a striking feature for those staying far away from the city centre since they have been spending more than 20% of their income on transport (Mudawu 2021). The unaffordability of transport leads to restricted accessibility that may drive an individual into poverty. As an adaptation mechanism, the urban poor have resorted to various unconventional modes of public transport such as riding on lorries and trucks that charge cheaper fares. Sustainable urban mobility should take into account inclusion and the promotion of non-motorised transport such as walking and cycling.

7.2.4 Central Government Intervention in Urban Roads

It can be noted that, due to increased urbanisation in Zimbabwe's urban spaces, there has been an increase in planned and unplanned settlements in peri-urban areas (Matsa et al. 2021). This has increased transport demand which the formal system has failed to cover (Government of Zimbabwe 2012; Dumba 2017). Urban development has focused on construction of buildings rather than the road system, resulting in dilapidated urban roads and potholes that have become a menace. To improve traffic circulation and mobility on Zimbabwe's urban roads, the government implemented the Emergency Road Rehabilitation Programme (ERRP) that runs for 36 months and is divided into three phases with US $400 million raised through ZINARA and Treasury (Mugabe 2021). The programme is mainly for improving urban roads for sustainable transportation, accessibility, and mobility in line with the Sustainable Development Goals (SDGs), particularly SDG 11 and the New Urban Agenda.

8 Conclusion and Recommendations

Transportation, mobility and accessibility challenges in Zimbabwe's cities mostly emanate from the monocentric nature of human settlements. The challenges have been exacerbated by land use zoning, low density development and rapid urbanisation which have resulted in urban sprawl. With increased car ownership by commuters, traffic congestion has been a major source of travel delay and environmental pollution. The government-sponsored public transport service (ZUPCO) has been inadequate and unreliable, leaving commuters to rely on own cars or informal transport operators for mobility and accessibility.

Zimbabwean cities were designed on the assumption of motorised transport, but with increasing urban poverty and congestion, the mode has become expensive thereby reducing the accessibility of destinations. In the case of Harare, it was found that the public transport system, which is heavily reliant on informal service providers, is not reliable.

The New Urban Agenda can contribute to sustainable urban mobility through restructuring the urban form to strengthen the linkage between land use planning and transport planning. Urban design should increase density so as to achieve economies of scale, but also mix land uses in order to reduce the need for automobility. The optimisation of urban density enables non-motorised modes of travel that make use of walking and cycling. The development of a national urban policy would help the articulation of the linkage between land use and transport for sustainable mobility in all urban settlements.

References

African Development Bank (2019) Zimbabwe Infrastructure Report 2019. African Development Bank Group. https://www.afdb.org/fileadmin/uploads/afdb/Documents/Project-and-Operations/Zimbabwe_Infrastructure_Report_2019_-_AfDB.pdf. Accessed on 09/09/2021

Baindur D, Rao P (2016) Equity in public transport—a case of Bangalore's city bus transport. J Sustain Urbanization, Plan Prog 1(1):44–55

Bandauko E, Bobo T, Mandisvika G (2016) Towards smart urban transportation system in Harare, Zimbabwe. In: Information Resources Management Association (IMRA), intelligent transportation and planning: breakthroughs in research and practice, vol 1. IGI Global, pp 962–978

Batty M, Marshall S (2017) Thinking organic, acting civic: the paradox of planning for cities in evolution. Landsc Urban Plan 166:4–14

Behruz H, Chavoshy AP, Mozaffari G (2013) Challenges of implementation of intelligent transportation systems in developing countries: case study–Tehran. WIT Trans Ecol Environ 179:977–987

Cervero R (2013) Linking urban transport and land use in developing countries. J Transp Land Use 6(1):7–24

Chikengezha T, Thebe V (2022) Living on the periphery and challenges of mobility: a tale of transport-induced social exclusion in Southlea Park, Harare, Zimbabwe. Urban Forum, pp 267–279

Dumba S, Vassileva LD, Gumbo T (2017) Methodological issues in modelling signalised intersection capacity under informal public transport operations: case study, Harare, Zimbabwe. Transp Res Procedia 25:4891–4915

Dumba S (2017) Modelling signalised intersections' capacity under the impact of minibus public transport in Harare, Zimbabwe. Doctoral Dissertation submitted to the University of Zimbabwe, Harare

Gkoumas K, Marques dos Santos FL, Stepniak M, Pekár F (2021) Research and innovation supporting the European sustainable and smart mobility strategy: a technology perspective from recent European union projects. Appl Sci 11(24):11981

Government of Zimbabwe (2012) National transport policy 'revitalising the transport sector in order to make the economy tick'. Ministry of Transport, Communication and Infrastructural Development, Harare

Government of Zimbabwe (2020) National Development Strategy 1 (January 2021–December 2025) Towards a prosperous & empowered upper middle-income society by 2030. Government Printers, Harare

Gryparis E, Papadopoulos P, Leligou HC, Psomopoulos CS (2020) Electricity demand and carbon emission in power generation under high penetration of electric vehicles. A European Union Perspective. Energy Rep 6:475–486

Klar S, Leeper TJ (2019) Identities and intersectionality: a case for purposive sampling in survey-experimental research. In: Experimental methods in survey research: techniques that combine random sampling with random assignment, pp 419–433

Kumar A, Zimmerman S, Arroyo Arroyo F (2021) Myths and realities of informal public transport in developing countries

Kundu D, Pandey AK (2020) World urbanisation: trends and patterns. In: Developing national urban policies: ways forward to green and smart cities, pp 13–49

Leeson GW (2018) The growth, ageing and urbanisation of our world. J Popul Ageing 11:107–115

Litman T, Steele R (2017) Land use impacts on transport. Victoria Transport Policy Institute, Victoria, Canada

Liu H, Fang C, Miao Y, Ma H, Zhang Q, Zhou Q (2018) Spatio-temporal evolution of population and urbanization in the countries along the Belt and Road 1950–2050. J Geog Sci 28:919–936

Marondedze AK, Schütt B (2019) Dynamics of land use and land cover changes in Harare, Zimbabwe: a case study on the linkage between drivers and the axis of urban expansion. Land 8(10):155

Mashingaidze N, Chirisa I, Mutambisi T, Matamanda AR (2021) Paradoxes surrounding carbon credits and local area development: the case of Mbire District, Zimbabwe. Local Environ 26(10):1175–1185

Matenga M (2021) Cabinet blames fuel availability for congestion. In: Newsday Newspaper (12 May 2021). Alpha Media Holdings, Harare

Matsa M, Mupepi O, Musasa T (2021) Spatio-temporal analysis of urban area expansion in Zimbabwe between 1990 and 2020: the case of Gweru city. Environ Challenges 4:100–141

Mbara T, Dumba S (2017) An analysis on institutional preparedness to bus route tendering in the context of a fragmented liberalized market: case study of Harare, Zimbabwe. https://ses.library.usyd.edu.au/bitstream/handle/2123/17416/Thredbo_15_Thredbo_15_Paper_15.pdf?isAllowed=y&sequence=1. Accessed on 13/02/2022

Mbara T, Pisa N (2019) An analysis of impediments to deliver sustainable transport in cities of developing countries: the case of Harare, Zimbabwe. WIT Trans Built Environ 182:241–252

Mbara TC (2006) Coping with demand for urban passenger transport in Zimbabwe: challenges and options. In: SATC 2006

McCann P (2015) The regional and urban policy of the European Union: cohesion, results-orientation and smart specialisation. Edward Elgar Publishing

Middleton L (2018) Freedom to move: transportation in African cities. AURI Policy Brief 6, 2018, p 13. Available online: https://www.urbanafrica.net/wp-content/uploads.../Mobility_ENG.pdf. Accessed 22 Dec 2022

Mudawu A (2021) Intergrating land-use and transport planning for easy mobility towards a sustainable transport planning system in Harare. Women University In Africa, Harare

Mugabe D (2021) Massive road rehabilitation on cards, The Sunday Mail 21 Feb 2021. https://www.sundaymail.co.zw/massive-road-rehab-on-cards. Accessed 22 Dec 2022

Mupfumira P, Wirjodirdjo B (2015) An economically sustainable urban public transport framework: system dynamics modelling approach to reduce public transport liberalisation impact in Harare City. Int J Sci, Technol Soc. Special Issue: Design Innovations for Sustainable Development 3(2–2):11–21. https://doi.org/10.11648/j.ijsts.s.2015030202.13

Mweshi GK, Sakyi K (2020) Application of sampling methods for the research design. Arch Bus Rev 8(11):180–193

Nyatondo T (2013) Determinants of the supply of urban public transport services in Harare, Zimbabwe (Master's thesis, University of Cape Town), Cape Town, South Africa

Palop BA (2020) Local leadership and its limits in the deployment of sustainable mobility policies. In: Finck M, Lamping M, Moscon V, Richter H (eds) Smart urban mobility. Springer, Berlin, Heidelberg, pp 81–98

Peeters P, Gössling S, Klijs J, Milano C, Novelli M, Dijkmans C, Eijgelaar E, Hartman S, Heslinga J, Isaac R, Mitas O (2021) Research for TRAN Committee-overtourism: impact and possible policy responses

Sexton MG (2000) Sustainable built environments and construction activity through dynamic research agendas. University of Salford (United Kingdom)

The Chronicle (2020) Transport blues as ZUPCO fails to cope with demand, 8 May 2020. https://www.chronicle.co.zw/transport-blues-as-zupco-fails-to-cope-with-demand/. Accessed on 19-11-2020

Tichagwa CG (2016) Unlicensed taxis in Zimbabwe's urban areas: the case for legalising an informal urban transportation system. Dev South Afr 33(1):81–98

Turok I (2015) Turning the tide? The emergence of national urban policies in Africa. J Contemp Afr Stud 33(3):348–369

UN-Habitat (2016) World cities report 2016: urbanization and development-emerging futures. UN-Habitat, Nairobi

UN-Habitat (2017) Action framework for implementation of the new urban agenda. UN Habitat, Nairobi

UN-Habitat (2020) The new urban agenda illustrated. Nairobi, UN Habitat. https://unhabitat.org/the-new-urban-agenda-illustrated. Accessed on 15/09/2021

United Nations (2017) New urban agenda, resolution adopted by the General Assembly on 23 December, 2016, New York, USA

Watson I, Ali A, Bayyati A (2019) Freight transport using high-speed railways. Int J Transp Dev Integr 3(2):103–116

Zeeman AS, Booysen MJ (2013) Combining speed and acceleration to detect reckless driving in the informal public transport industry. In: 16th International IEEE conference on intelligent transportation systems (ITSC 2013). IEEE, pp 756–761

Zikhali W (2017) Potholes in Zimbabwe: a hindrance to economic development. Eur J Res Soc Sci 5(2):59–70

Zimbabwe National Statistics Agency (ZimStat) (2022) Census 2022: provincial report Harare. ZimStat, Harare

Zimeye (2021) Inter-ministerial team to deal with congestion in Harare. Accessed 26/07/21 from inter-ministerial team to deal with congestion in Harare—ZimEye

Quantity Surveying Steering the New Urban Agenda in Zimbabwe

Tirivavi Moyo and Benviolent Chigara

Abstract Sustainability learning deficits are evident in the quantity surveying profession in Zimbabwe, particularly regarding achieving the New Urban Agenda through provision of robust and resilient infrastructure. However, the derivation and implementation of remedies still lag. Therefore, this study aims to explore strategies that can be implemented for quantity surveyors to effectively contribute to the New Urban Agenda. Structured interviews of key informants (quantity surveying academics, the quantity surveying professional body educational committee representatives, and relevant government departments heads) were utilised to collect qualitative data. Thematic analysis revealed key learning and work-related fundamentals. These include expanding the curriculum for under and postgraduate quantity surveying degree programmes to include sustainable cities and resilient infrastructure aspects. Further, the need for sustainable cities continuous professional development programmes for quantity surveying professionals is supported. Holistically, all construction professionals need to improve their sustainable development and construction literacy. Recommendations include the promulgation of statutory stipulations that support robust sustainability-enhancing construction projects. Views of quantity surveying professionals were emphasised as these are best suited to reveal their profession's inadequacies. Social and environmental sustainability literate quantity surveyors and construction economists are paramount to the realisation of the New Urban Agenda through relevant cost advice and alternative competitive solutions.

Keywords Quantity surveyors · Construction economics · New urban agenda · Sustainable cities · Sustainability literacy

---

T. Moyo (✉)
Department of Quantity Surveying, Nelson Mandela University, Summerstrand North, Gqeberha, South Africa
e-mail: tirivavimoyo@gmail.com

B. Chigara
Faculty of the Built Environment, National University of Science and Technology, Bulawayo, Zimbabwe

C. Chavunduka and I. Chirisa (eds.), *New Urban Agenda in Zimbabwe*,
Advances in 21st Century Human Settlements,
https://doi.org/10.1007/978-981-97-3199-2_9

1 Introduction

The promotion of urban sustainability provides an opportunity to rethink urban planning and development (Valencia et al. 2019). This is paramount for both developed and developing countries due to the overwhelming environmental, economic, and social issues within their cities. Thus, the promulgation of the New Urban Agenda (NUA) in 2016 was momentous as it was critical (United Nations Economic Commission for Africa 2020). Suffice to say, the agenda sets to attain the urban sustainability goals within the United Nations 2030 developmental framework (Garschagen et al. 2018). In Zimbabwe, challenges of inadequate housing, inadequate provision of services, and environmental pollution (Munzwa and Wellington 2010) require a sustainability-entrenched response from the built environment professionals. Thus, solutions should be consequently borne within the increasing concept of sustainability and equipping professionals in its literacy (Murray and Cotgrave 2007; Presley and Meade 2010).

While there should be numerous built environment professionals that should play significant roles in the fulfilment of the NUA, this study focused on quantity surveyors. In Commonwealth countries, the term quantity surveyor is popular; however, they are also known as construction economists in Continental Europe and cost engineers in North America (International Federation of Surveyors 2018). Quantity surveyors are critical for the financial and contractual management of construction projects (Pacific Association of Quantity Surveyors 2001; Shayan et al. 2019) in towns and cities. Quantity surveyors in Zimbabwe, like in most developing countries, have mainly focused on such traditional competencies (Seah 2009; Omotayo et al. 2022) and have not aligned these to the fulfilment of urban sustainability development goals. This is despite, Chirisa (2014) reporting that cost issues play a significant role in shaping urban environments, and these relate to the work of quantity surveyors. Further, Chamikara et al. (2020) advocate for these professionals to enhance their sustainable construction knowledge as it has become essential. Hence, the urban sustainability goals outlined in the NUA require the input of sustainability competent quantity surveyors. Since sustainability and innovation challenges are pre-dominant in the Zimbabwe construction industry (Mhlanga 2018), there is need to explore how these professionals can be capacitated to respond to achieving the sustainable cities' goal. Therefore, this study sought to explore strategies to steering the NUA agenda through incorporating the competencies of the quantity surveyor. There is need for quantity surveyors to progressively develop their skills and knowledge in order to deliver effective cost-advisory services for green projects (Ma and Luu 2013). Important areas for skills upgrading may include green costing and carbon cost planning for sustainable buildings. In pursuance of this, Piercy and Rich (2015) also support that sustainability should be driven by dimensions of community, governance, and quality issues, amongst others. Also, successful sustainable delivery of construction projects is paramount (Agyekum-Mensah et al. 2012). Cumulatively, training and skills development is crucial in promoting attitudinal change in environmental awareness, social progression, and economic development (Cotgrave and Kokkarinen 2010) for the realisation of sustainable cities and communities.

The next section of the chapter presents the theoretical framework of the study, which includes the NUA and quantity surveying nexus. Thereafter, the methodology used to address the research questions is explained and the findings are outlined. Finally, the conclusions on the resolutions towards steering the NUA are outlined.

2 Literature Review

This section considers the theoretical framework of the NUA, education for sustainable development and the quantity surveyors' current standing within the NUA drive.

2.1 New Urban Agenda

The New Urban Agenda (NUA) is a development framework that seeks to promote sustainable urban development. The NUA was adopted as a development framework for urban areas by the United Nations Conference on Housing and Sustainable Development (HABITAT III) in Quito, Ecuador, on 20 October 2016 (UN-HABITAT 2017). The vision of the NUA is to create cities for all, seeking to promote inclusivity and ensure that all inhabitants, of present and future generations, without discrimination of any kind, are able to inhabit and produce just, safe, healthy, accessible, affordable, resilient, and sustainable cities and human settlements to foster prosperity and equality for all (UN-HABITAT 2017).

The adoption of the NUA is premised on two main factors—the opportunities and challenges presented by rapid urbanisation. Urban population is estimated to double by 2050 (UN-HABITAT 2017) with approximately 2.5 billion people being added to urban areas (World Economic Forum, WEF 2017). Urbanisation presents several opportunities to stimulate economic growth, reduce poverty, and contribute to sustainable urban development (Clos 2017). Thus, in the NUA, cities are seen as enablers of the sustainable development agenda rather than challenges / problems. According to the WEF (2017), urbanisation is an essential part of growth and a key factor in socio-economic transformation of cities. Nonetheless, accelerated urbanisation contributes to environmental, economic, and social challenges in cities. Cities face multiple forms of challenges such as poverty, growing inequalities and environmental degradation, social and economic exclusion, spatial segregation, housing affordability, climate change, rising informal settlements, traffic congestion, lack of administrative, and financial capacity (UN-HABITAT 2017; WEF 2017). These problems have several ramifications relative to the realisation of the Sustainable Development Goal number 11—making cities and human settlements inclusive, safe, resilient, and sustainable. The NUA seeks to address existing and new challenges of cities through coming up with sustainable, inclusive, and resilient strategies to address the challenges experienced by cities. The thrust of the SDGs is to end poverty, protect the planet, and ensure prosperity for all (Watson 2016). The NUA embodies three principles, namely "leave no one behind, by ending poverty in all its forms and dimensions, including eradication of extreme poverty"; "ensure sustainable and inclusive urban economies and opportunities for all"; and "ensure environmental sustainability (WEF 2017)". The NUA lays out the standards and principles for the planning, development, management, and improvement of urban areas (Clos 2017).

To realise the NUA goal, collective effort of various stakeholders is required. As most cities lack the capacity and resources to ensure that the city develops in a sustainable manner, building the transformative strategies to positively shape urbanisation outcomes calls for multi-stakeholder cooperation (Charles and Buchler 2017). While the public sector stakeholders—national and sub-national authorities—are identified in the agenda as primary stakeholders in the provision of urban infrastructure and services (UN-HABITAT 2017), a collaborative approach, which creates space for the private sector, civil society, and academia to contribute to the realisation of the agenda at all stages of the urban value chain is required (WEF 2017). According to the WEF (2017), universities provide cities with talent that can develop the economic character of the city, and specialised expertise in urban domain can be leveraged to contribute to urban design and planning. Watson (2016) acknowledges the recognition of planning experts in the implementation of the NUA. The aforesaid is corroborated by Caprottia et al. (2017) who stress the importance of experts and their implications for the production of specific visions of what can be considered as the "new urban citizen".

Valencia et al. (2019) observed that while the NUA and SDGs agendas are "ambitious" and probably not achievable within the timeframe, the frameworks provide an opportunity to rethink planning and development with all the three dimensions of sustainability in mind. Further to this, reporting on the initial results of a comparative study to monitor implementation of SDG 11 alongside the NUA in seven cities on four continents, namely Buenos Aires, Cape Town, Gothenburg, Kisumu, Malmö, Sheffield, and Shimla, the study highlights five key aspects important when implementing the NUA and SDGs: delimiting the urban boundary, integrated governance, actors, trade-offs and synergies, and indicators. Nonetheless, despite identifying the importance of multi-stakeholders in implementing the SDGs and NUA, except for planners, the role of other built environment professionals is conspicuous by its absence in the narrative of the actors. Of these built environment professionals, quantity surveyors are critical to NUA's achievement; however, little focus seems to have been directed towards them.

2.2 *Education for Sustainable Development*

Pappas (2012) echoes that the deteriorating environmental, economic, and social circumstances necessitate an answer from academics. The sustainability concept seeks to integrate stakeholders' concerns to ensure basic needs, ecology, eco-efficiency, and community empowerment are achieved (Virakul 2015). Gu et al. (2011) suggest barriers towards coalescing sustainable development in technical education and training as including its misunderstanding. Despite the authenticity of this challenge, United Nations Education Science and Cultural Organisation (UNESCO) (2014) outlines one of its guiding principles as "Education for sustainable development (ESD)". Further to this, one of its 2030 targets is to ensure all learners acquire knowledge, skills, values, and attitudes to establish sustainable and peaceful societies, through global citizenship education and education for sustainable development. Gu et al. (2011) opine the worth of ESD as a theoretical and practical basis to usher society towards sustainability.

The inexactness, contextual uniqueness, and ignorance of sustainability are justified as the stumbling block of mainstreaming sustainability (Blewitt 2010). However, Viertel (2010) laments arguments on prevalent inequalities, environmental damage, dwindling resources, and the consequent need for stronger technical education on sustainable development. Cruickshank and Fenner (2012) admit that contemporary societies increasingly need elaborate socio-technical systems to deliver adequate and compliant products and services. To counter these challenges, UNESCO (2014) supports the strengthening of ESD through implementing considerable changes in training which entail integrating critical issues such as sustainable consumption and production. Also, UNESCO (2012) notes that ESD can develop learning objectives that are specific to a particular sustainable development goal (SDG) and relevant to all 17 sustainable development goals.

The International Labour Organisation (ILO) (2011) asserts that to keep training relevant, institutional and financial arrangements must bridge the world of learning and the world of work through bringing business, labour, government, and training providers together. Quality sustainable education is now more relevant than ever. Kronner (2005) also report that contextualisation, application, judgement, and decision-making are vital for embedding sustainable development within our professional working activities, and these can only be achieved through quality sustainable education. Viertel (2010) concurs that the capable evolution to attaining sustainability is vital and can be realised through general awareness, applied research, technological innovation, and provision of advice and assistance to individuals and companies. Further, Viertel (2010) proffers key competencies of environmental awareness and knowledge, technical and business knowledge, and skills as being fundamental for technical training for achieving sustainable development. Also, ILO (2011) emphasises that great effort is required to ensure skills training systems deliver quality training through an adequate supply of qualified training staff.

2.3 *Quantity Surveying and the New Urban Agenda*

The Quantity Surveyors Act Chapter 27:13 (2001) was first promulgated in 1981 to "establish a council for the profession of quantity surveying to be known as the Quantity Surveyors Council of Zimbabwe and to provide for its functions and powers; to provide for the registration and regulation of the practice of quantity surveyors in Zimbabwe; and to provide for matters incidental to or connected with the foregoing". The Act situates the statutory requirements for the functions of quantity surveyors in Zimbabwe. Quantity surveying tertiary education is undertaken by five (5) polytechnic colleges and two universities. The Zimbabwe Institute of Quantity Surveyors' (ZIQS), which exists within the domain of the Quantity Surveyors Council of Zimbabwe, is the professional body responsible for technician and professional membership and ensures competency testing within the profession. The ZIQS competency testing framework addresses five main knowledge areas, namely measurement and estimating, professional practice, professional fees, contracts administration, and final accounts. Situated within these knowledge areas are competencies that include contract practice (due diligence); construction technology and environmental services (constructability and the environment); procurement and tendering (general procurement advice, and tendering process); project financial control and reporting; quantification and costing of construction works (cost estimating, measurement and statistical analysis); design economics and cost planning. These have been the expected major competencies to date. However, the absence of adequate research has resulted in the stagnation of the profession in the light of worldwide current trends, and this has contributed to current construction industry challenges (Mhlanga 2018; Moyo et al. 2019).

The quantity surveying profession occupies an important position in the realisation of the NUA. According to WEF (2017: 8), NUA is committed to (a) readdressing the way we plan, finance, develop, govern, and manage cities and human settlements; (b) recognising the leading role of national governments, as appropriate, in the definition and implementation of inclusive and effective urban policies and legislation for sustainable urban development; (c) adopting sustainable, people-centred, age- and gender-responsive, and integrated approaches to urban and territorial development. In summary, the NUA seeks to create sustainable and inclusive urban built environment. The contribution of quantity surveyors to these principles cannot be overstated. Seah (2009) believes that quantity surveyors are well positioned to lead in the sustainable age with cost-effective sustainable strategies coupled with well-placed value management and engineering contributions not only to bring down the green capital cost but also to personify the ethos of life cycle management, property performance appraisal, green financing, and strategies. This is corroborated by Omotayo et al. (2022) who reported that quantity surveyors contribute to the sustainable development of built urban environment through four main focal areas, namely sustainable construction, cost management, operation, and decommissioning. For example, NUA would require adequate and sustainable infrastructure for its needs to be fulfilled and quantity surveyors do play a significant role in realising such objectives through the competencies of construction technology and environmental services, and procurement and tendering. Also, the environmental, economic, and social challenges in cities are key issues that are at the centre of the quantity surveyor's roles and responsibilities. In addition, contracts administration of private–public partnerships financing for infrastructure development is critical contribution of quantity surveyors, and these are attained through their competencies of contract practice, procurement and tendering. With regards to the commitments, there is a considerable implication to the expected and existent roles and responsibilities of quantity surveyors especially in the aspect of planning, financing, developing cities, and human settlements.

Evidently, the NUA is constituted of aspects that would require the input of the quantity surveyor. Perera et al. (2007) opined that quantity surveyors have and should continue exploring opportunities beyond the traditional construction industry due to their multi-disciplinary competencies. Chamikara et al. (2020) also reported that the quantity surveyor's skills on sustainable construction are inherent in all of their essential competencies. Despite these relationships, limited studies exist on the role of the quantity surveyor in the NUA drive. However, numerous studies have acknowledged that the role of the quantity surveyor has evolved and continues to evolve (Crafford and Smallwood 2007; Dada and Jagboro 2012; Dada 2017; Akinola 2019; Shayan et al. 2019). To contribute to the NUA, quantity surveyors need to transform their services and offer cost advice on sustainable solutions and enhance flexibility in procurement management, risk, and value management and cost engineering (Omotayo et al. 2022). Thus, quantity surveyors need to be equipped with skills and expertise to provide the latest innovation application, green cost consultancy, cost-effective, sustainable strategies, building material solutions, life cycle cost analysis, and property performance assessment (Omotayo et al. 2022). This highlights the need for immediate attention in curricula design and continuous professional development to keep quantity surveyors relevant for the task ahead. In addition, despite the dearth of studies investigating the role of quantity surveyors in NUA, Seah (2009) and Omotayo et al. (2022) reveal that the traditional competencies of quantity surveyors are inadequate to contribute meaningfully to the sustainable built environment. Mandatory competencies of quantity surveyors, as outlined by Pacific Association of Quantity Surveyors (PAQS) (2001), Ramus et al. (2008), Australian Institute of Quantity Surveyors (2012), International Federation of Surveyors (IFS) (2018), and the Royal Institute of Chartered Surveyors (RICS) (2018), are devoid of specific aspects of the NUA concept. These include ethics, rules of conduct and professionalism; client care; communication and negotiation; data management (cost information database); diversity inclusion and team working; and health and safety. However, the core and optional/specialist competencies of design economics and cost planning, value management and engineering, sustainability issues, property investment development, project management, project feasibility analysis, research and development, amongst others, have a strong relationship to the NUA concepts. Thus, a need exists for a distinct revelation of the key aspects that would enable quantity surveyors to effectively contribute to the NUA. Yet before identifying the key aspects, exploration of strategies to enhance the inclusion of quantity surveyors in the NUA is pertinent.

3 Methodology

An interpretative philosophy was used due to the exploratory nature and lack of clear existing concepts in the reported study (Saunders et al. 2016). A qualitative inquiry was thus instituted for the study. Leedy and Ormrod (2015) support the use of interviews for such an inquiry. Hence, data was collected utilising structured interviews of key informants that included quantity surveying academics (three), a quantity surveying professional body educational committee representative (one), consulting quantity surveyors (two), and relevant government departments' heads (two), each from the Ministry of Local Government, Public Works and National Housing, and the Ministry of Transport, Communication, and Infrastructural Development. Thus, a total of eight (8) key informants were interviewed. Their profile is shown in Table 1. The respondents had a minimum experience of ten (10) years and a maximum experience of thirty (30) years in the Zimbabwean construction industry. Due to the subject area being uncommon, amongst quantity surveyors, purposive sampling was utilised (Babbie and Mouton 2015).

Table 1 Profile of respondents

Respondent	Designation	Highest qualification	Years' experience
R1	University Lecturer	M.Sc.	10
R2	University Lecturer	M.Sc.	10
R3	Polytechnic Lecturer	M.Sc.	12
R4	Polytechnic Lecturer	B.Sc. (Honours)	11
R5	Consultant Quantity Surveyor	M.Sc.	30
R6	Consultant Quantity Surveyor—Member of the Educational Committee of the Council of Quantity Surveyors in Zimbabwe	M.Sc.	12
R7	Chief Quantity Surveyor—Ministry of Transport	B.Sc. (Honours)	12
R8	Chief Director—Housing Development and Social Amenities	M.Sc.	20

3.1 Instrument Design and Administration

The structured interview guide consisted of five questions. The first question (i) considered the demographic profile of the respondents. This profile included the designations, qualifications, and experience of the informants within the quantity surveying profession. The other questions were ordered as follows: (ii) the respondents' understanding of the NUA, (iii) the current preparedness of quantity surveyors to contribute to the NUA, (iv) the anticipated challenges towards achieving the NUA from the quantity surveying perspective, and (v) the strategies to enhance the contribution of quantity surveyors to the NUA. The interviews were conducted telephonically and/or through the Microsoft teams' platform. The interview sessions were also recorded and transcribed. Prior to the interviews, the questions were pilot tested on academics within the quantity surveying professions and adjustments were made to the quality of the questions. This enhanced the validity of the interviews. Further, the use of various designations within the quantity surveying profession enhanced the validity of the study.

3.2 Data Analysis

Saunders et al. (2016) state that qualitative data is predominantly non-numeric and cannot be quantified. Naoum (2013) alludes that there are many techniques of analysing qualitative data and that qualitative data analysis is more difficult as compared to quantitative analysis. The analysis was aided by the use of ATLAS.ti version 8.0, where keywords and phrases were coded with predetermined responses/themes. Reliability of the data analysis was ensured using two coders and the ATLAS.ti version 8. Krippendorff (2004) suggests that the reliability of qualitative data is assured if the coders follow explicit and communicable instructions and work independently from each other. The validity of data was ensured by considering how the available evidence and theory support the findings. Elo and Kyngas (2007) relay that the credibility of research findings is strengthened by the resemblance of the codes to theory. Therefore, empirical validity was utilised in the study. Thematic analysis was utilised due to its flexibility in exploring different interpretations (Saunders et al. 2016) of the contributions of quantity surveyors to the NUA. The themes were linked to existing theories and were linked to the questions that formed the structured interview guide.

4 Discussion of Findings

In this section, the findings within the thematic areas of understanding of the NUA, preparedness of quantity surveyors to contribute to the NUA, challenges towards achieving the NUA, and strategies to enhance the contribution of quantity surveyors to the NUA are discussed.

4.1 Understanding the New Urban Agenda

The promulgation of the NUA in Zimbabwe was not reciprocated with a deliberate drive to capacitate all construction professionals with its concepts. This is despite the challenges presented by rapid urbanisation (UN-HABITAT 2017). All respondents acknowledged the importance of the NUA and the potential of quantity surveyors effectively contributing to its achievement through their various roles, thus confirming the need for a collaborative approach for realising this agenda (WEF 2017). Although professionals in property and planning-related studies seem to have made strides in this agenda, quantity surveyors still lag. Respondents from the academic institutions (universities and Polytechnics) and the education committee of the quantity surveying council of Zimbabwe, all concur that quantity surveyors lack the requisite understanding and appreciation of the NUA to significantly contribute to it as exemplified by respondent 3 and respondent 6's comments.

R3 "*these issues are yet to be incorporated into the curriculum of Quantity surveying technicians, hence they are still ill prepared in that respect*".

R6 "*most Quantity surveyors do not have the knowledge or appreciation of the New Urban Agenda*".

However, quantity surveyors have a different perspective. They purport that they are ready to contribute to the needs of the NUA, despite not undertaking any continuous professional development programmes that address those concepts. Their argument was that the profession is dynamic and the roles of the quantity surveyor can easily align with the dictates of the originators of construction projects. Hence, any conceptual changes by clients would be effectively addressed by quantity surveying professions, as alluded to by respondent 5. Further, they reported that any needs would be dealt with through research.

R5 "*due to the New Urban Agenda being development related, the Quantity surveyors guide financial and cost aspects of such construction projects within the agreed briefs*".

On the other hand, respondents R 7(as expanded by respondent 7) and R8 from government departments agreed with the sentiments of academics and alluded to the need to instruct Quantity surveyors within their departments concerning the NUA.

R7 "*the New Urban Agenda is a new concept and we are not yet well versed with it*".

R8 "*It is highly likely that quantity surveyors have not been exposed to the concepts of the NUA*".

The study results corroborate studies conducted by Bolade-Oladepo et al. (2020) that the effectiveness of the quantity surveying profession in ensuring sustainability in the built environment is affected by limited awareness in social responsibility and limited awareness in sustainable construction. Despite the varied views from the construction professionals, it is apparent that more needs to be done to enhance the understanding of the NUA. Valencia et al. (2019)'s observation that the agenda is not achievable within the timeframe is justified by the lack of widespread awareness of the concept by quantity surveyors. The training of quantity surveyors at polytechnic and university levels should be recalibrated to ensure that graduates are well instructed on the NUA issues. Suffice to say, the existing learning objectives require recalibration to ensure such NUA objectives are incorporated as alluded to by WEF (2017). Although the quantity surveying practitioners contend that they are well equipped through research to tackle the issues, there is no evidence to support this assertion. Research output from these professionals on the NUA is non-existent. Hence, continuous professional development programmes, administered by property, and planning professionals, are essential to ensure that the practitioners are conversant with the issues. Beyond these measures, in most instances, governmental policies have been effective in steering agendas. Thus, policies that enforce professional development in these aspects are supported.

4.2 Preparedness of Quantity Surveyors to Contribute to the New Urban Agenda

From the earlier discussion on the understanding of the NUA by quantity surveyors, it is apparent that their preparedness to contribute to its achievement is inadequate. This is potentially linked to its misunderstanding as reported by Gu et al. (2011). Academics and government departmental heads are convinced that it is paramount for quantity surveyors to be capacitated in this concept. However, they also agree that there is a huge gap in sustainability development instruction, as alluded to by Blewitt (2010); Viertel, (2010); and Cruickshank and Fenner, (2012).This is constituted by the lack of impartation of sufficient NUA knowledge, for the quantity surveyors to be adequately prepared. Two of the respondents categorically stated that,

R3-*we are yet to adequately instruct the students for them to be sufficiently prepared to effectively contribute to the New Urban Agenda.*

R7-*we are currently far from being prepared to drive this agenda.*

The results reinforce the findings of Bolade-Oladepo et al. (2020) that quantity surveyors are not well equipped with knowledge and skills about sustainability and lack sufficient cost database for building services and products. However, in contrast to the other respondent groups, the quantity surveying practitioner (R5) was of the view that "*the professionals are ready to ensure the agenda is achieved and that they are capacitated enough*". It is a possibility that such practitioners have attained the requisite knowledge from their interactions with other built environment professionals as they implemented property development projects. Such a situation would simplify the permeation of the requisite knowledge to the other professionals. However, the lack of cohesion within the professional bodies, as espoused by R8, as highlighted in the next sub-section, strongly distracts any information sharing platforms. Contrariwise Virakul (2015) highlight a need for stakeholder integration to enhance such sustainability concepts. Having such a scenario also complicates any targeted interventions. The most logical course of action would be to ascertain those that are not prepared and institute interventions. Also, it would be pertinent to holistically steer all the professionals towards preparedness, as suggested by WEF (2017).

4.3 Challenges Towards Achieving the New Urban Agenda

The realisation of the NUA is plagued by a plethora of challenges ranging from lack of governmental policy interventions to lack of foresight for professional developmental initiatives within the quantity surveying professional body. All the respondents (R1 to R8) paint an insurmountable picture; however, most of the highlighted challenges can be overcome by an inclusive approach by all the construction stakeholders as supported by ILO (2011). Despite the lack of synergy (WEF 2017) within the various construction-related professional bodies being recognised as a significant challenge, the recent establishment of the Zimbabwe Construction Industry Council can go a long way in unifying the industry. The Zimbabwe Construction Industry Council is mandated through the approved Zimbabwe Construction Council Bill (2019) to align contractors and the built environment to international best practices. This will inevitably enable the industry to improve knowledge sharing with quantity surveyors benefiting from the other professions who are well capacitated on the NUA drive. Thus, strengthening education for sustainability (UNESCO 2014) and ensuring quality sustainable education (Kronner 2005) are vital. One of the experts from the governmental department highlighted that:

R8- *there is no synergy within the professional bodies, and this leads to a dearth in information transference which would have likely benefited Quantity surveyors.*

Further, the absence of an integrated construction Act has been identified as a major barrier. The existence of professional bodies and construction stakeholders as "silos" prevents the quantity surveying profession from benefiting from international best practices, like NUA, from related construction professionals. Consequentially, any positive strides on NUA are negated by critical financial and contractual management professionals who lack such basic principles of NUA. The existence of various acts (National Social Security Authority Act, Environment Management Agency Act, Procurement and Disposal of Public Assets Act) and statutory instruments, situated in various ministries, complicates their implementation and enforcement in the construction industry. Nationalisation of the NUA is essential to ensure policies reflect its goals (UN-HABITAT 2017). The Construction Industry Bill has been touted as a long-term solution to this problem. However, its inclusion of such aspects is unclear. In addition, the relevant Ministry responsible for local government, public works, and national housing has not reacted timely towards enabling professional bodies, which include the quantity surveying professional body, to capacitate themselves in the requirements of the NUA. The piecemeal approach by the Ministry has left an instructional gap within construction professionals. Hence, deliverables that include protecting the planet (Watson 2016) through sustainable consumption and production (UNESCO 2014) are increasingly unattainable.

The unavailability of quantity surveying instructors who are competent in the NUA has also been highlighted by academic experts (R1 to R4). There is a need for lecturers to be trained in the NUA and enable them to instruct it within the quantity surveying context. ILO (2011) supports the availing of adequately qualified training staff to ensure competent training systems. The lack of funded research within the academic fraternity on these aspects worsens the issue. Effort should be made within the construction industry to drive any national initiatives, including the NUA, through availing resources to enable research by academics and capacitation of construction professionals. This approach is supported by the WEF (2017) and will guarantee the talent that is needed to achieve sustainable cities and communities. Brain drain has also negatively affected any efforts to enhance the implementation of the NUA. The current economic environment is not suitable for construction professionals. This has culminated in the most experienced academics and professionals migrating to other countries within in the region and abroad. Such migration consequently has a negative effect within the Zimbabwean construction industry, as the inexperienced professionals struggle to implement such novel initiatives.

Some respondents have proffered sentiments that the NUA presents financing challenges as most of compliant developments would be costlier compared to the traditional projects. In an industry where funding of construction projects has already been problematic, this convolutes the problem. Viertel (2010) bemoans the dwindling resources, which should include financial resources, on addressing sustainable development concerns. In addition, the economic environment does not support the NUA due to clients opting for less expensive developments that are devoid of sustainability constructs. This is exacerbated by the presence of economic sanctions that restrict support from external investors. This situation is pertinent to quantity surveyors as they are the construction cost and financial experts (PAQS 2001; Ramus et al. 2008). Thus, the lack of capacity of quantity surveyors on the NUA is likely linked to an existing construction property development clientele which has been reluctant to implement its concepts.

4.4 Strategies to Enhance the Contribution of Quantity Surveyors to the New Urban Agenda

Resolutions towards enhancing the contribution of quantity surveyors to the NUA resonate from the more common strategies of having seminars and workshops to policy shifts. All the respondents (R1 to R8) agreed on these common strategies. Multi-stakeholder conferences, seminars, and workshops are paramount as they present a platform for research and knowledge sharing on these aspects, as alluded to by WEF (2017). Construction professionals of different designations can come together and plan for the implementation of the NUA in unison. The current fragmented approach should be a thing of the past and where there are inadequacies; locally, efforts should be made to collaborate with regional and international academics and professionals. This enables the various professional bodies, including that of quantity surveyors, to glean from other stakeholders and effectively contribute to the NUA. Such proactive strategies also culminate in establishment of continuous professional development programmes that are geared towards improving the knowledge of the NUA, amongst other future agendas.

It is paramount that the required learning objectives for quantity surveyors, to be effective in the implementation of NUA, be established. From the literature and perceptions of respondents, this is critical to resolving this gap. Emphasis on the social and environmental sustainability aspects within urbanisation should be primarily focused on. Although these are important aspects, care must be taken not to inculcate key competencies of other built environment professionals into those of quantity surveyors. These should just be sufficient for quantity surveyors to have an appreciation of the NUA objectives and for them to be able to interpret these objectives and successfully implement them. This will enable effective implementation of the technical and economic sustainability requirements of the NUA, which are synonymous with the roles and responsibilities of quantity surveyors. The evolution of quantity surveying competencies and the required consequent dynamism in their incorporation is also important as highlighted by (Crafford and Smallwood 2007; Dada and Jagboro 2012; Dada 2017; Akinola 2019; Shayan et al. 2019). This can present a policy shift, especially in the procurement of quantity surveying consultancy services where competence in the NUA concept should be made mandatory and where, if this lacks, joint ventures with capable regional and international firms are allowable. These perceptions were offered by R7 and R8, as they responded from a policy perspective. Despite Chamikara et al., (2020) reporting that sustainable construction skills are inherent in all competencies of quantity surveyors, and the evidence from respondents shows that much needs to be done. As these interventions are incorporated in a construction Act, a dynamic approach to incorporation of construction-related national and international concepts is established. In addition, incentives for incorporating NUA concepts will go a long way in stimulating the quantity surveying training institutions and professional body to effectively plan and incorporate it into curricula design and delivery on construction and infrastructural developmental projects. Tax exemptions are effective tools to encourage clients to develop sustainably and within the confines of the NUA. For example, tax incentives can be offered to property developers that incorporate NUA principles within their development concepts, and these can be achieved for procurement and maintenance costs of technology that supports the NUA. Further, penalties for those that produce and deliver projects that are not conversant with the NUA concept, are also implementable.

5 Conclusion and Recommendations

The NUA's prerogative is to ensure sustainable cities and communities are a reality, and this has become a national initiative for both developed and developing countries. With construction activities contributing to environmental pollution and property developmental projects lacking in provision of adequate services, it has become pertinent for built environment professionals to become significantly involved. Amongst these professionals, quantity surveyors should shoulder an important role of contributing to the economic, social, and environmental sustainability of projects through cost and contractual interventions. However, quantity surveyors in the Zimbabwean construction industry seem inadequately prepared to effectively contribute to this agenda. Therefore, this study sought to qualitatively explore resolutions to this challenge. The interviewed key informants generally agreed that quantity surveyors lack the adequate understanding to drive the NUA. However, quantity surveying practitioners indicated that they could easily transition to achieve the concept. While their positive assertion is appreciated, any considerable gain is founded in the successful integration of the profession, from training of graduates to professionals in the industry, being ably equipped. Hence, there is a need for an intensive and extensive deliberate instructional programme for students and practitioners. This was echoed by the apparent lack of preparedness of quantity surveyors to effectively contribute to this drive. A holistic and targeted intervention approach was suggested to ensure a comprehensive construct.

Numerous challenges were identified by the key informants. From lack of adequate policy to steer the agenda, specifically, within the construction industry to the general lack of synergy within built environment professional bodies. Also, the absence of a dynamic construction Act that would adequately respond to any needs within the construction industry was bemoaned. This culminated in the key informants suggesting strategies to enhance the contribution of quantity surveyors to the NUA. The most important strategies were focused on ensuring that a responsive construction policy be enacted that would be sensitive and would effectively respond to novel concepts that would affect the construction industry and incorporate aspects of research and development. An integrative and multi-disciplinary approach throughout the property development and project lifecycle is most suited to enhance the realisation of the NUA vision. This entails an extensive sustainability education drive through academic institutions and professional bodies and the integration of NUA-specific social and environmental sustainability aspects into various "functions" of quantity surveyors in the project lifecycle. These aspects should include learning of how planning, financing, developing, governing, and managing cities and human settlements can be undertaken. Also, adopting sustainable, people-centred, age- and gender-responsive, and integrated approaches to urban and territorial development by implementing policies, strategies, capacity development and actions at all levels, based on fundamental drivers of change. Regulatory authorities can also be proactive in shaping the sustainability agenda by insisting on adherence to sustainability enhancement proposals. The study had limitations as it did not consider other professionals in the built environment; however, due to its exploratory nature, the quantity surveyors' views were considered to be sufficient and valid. Future studies can consider both quantitative and qualitative methodologies to investigate the specific needs within the training of students and continuous professional development of quantity surveying professionals.

References

Agyekum-Mensah G, Knight A, Coffey C (2012) 4Es and 4 Poles model of sustainability. Struct Surv 30(5):426–442. https://doi.org/10.1108/02630801211288206

Akinola JA (2019) Challenges of adopting multiskilling strategy to quantity surveying practice in Nigeria. IOSR J Mech Civ Eng. 16(6):58–66

Australian Institute of Quantity Surveyors (2012) National competency standards for quantity surveyors—construction economists (ISBN 1-876389-02-8). The Australian Institute of Quantity Surveyors, Sydney

Babbie E, Mouton J (2015) The practice of social research. South African Edition. Contributions by Vorster P, Prozesky B Oxford University Press, Southern Africa. Cape Town

Blewitt J (2010) Higher education for a sustainable world. Educ + Train 52(6/7):477–488. doi:https://doi.org/10.1108/00400911011068432

Bolade-Oladepo FA, Oladipo FO, Fasuyi OA (2020) Sustainable built environment: the role of quantity surveyors. J Eng 10(1):01–06

Caprottia F, Cowley R, Datta A, Broto VC, Gao E, Georgeson L, Herrick C, Odendaal N, Joss (2017) The new urban agenda: key opportunities and challenges for policy and practice. Urban Res Pract 10(3):367–378. https://doi.org/10.1080/17535069.2016.1275618

Chamikara PBS, Perera BAKS, Rodrigo MNN (2020) Competencies of the quantity surveyor in performing for sustainable construction. Int J Constr Manag 20(3):237–251

Charles A, Buehler M (2017) Foreword. In: World Economic Forum (2017) Harnessing public-private cooperation to deliver the new urban agenda. World Economic Forum, 1–9

Chirisa I (2014) Building and Urban Planning in Zimbabwe with special reference to Harare: putting needs, costs and sustainability in Focus. Consilience: J Sustain Dev 11(1):1–26

Clos J (2017) Foreword. In: UN-HABITAT (2017) New urban agenda. UN-HABITAT, Nairobi, iii–v

Cotgrave AJ, Kokkarinen N (2010) Developing a model promoting sustainability literacy through construction curriculum design. Struct Surv 28(4):266–280. https://doi.org/10.1108/02630801011070975

Crafford GJ, Smallwood JJ (2007) Clients' views on quantity surveying competencies. Acta Structilia: J Phys Dev Sci 14(1):33–55

Cruickshank H, Fenner R (2012) Exploring key sustainable development themes through learning activities. Int J Sustain High Educ 13(3):249–262

Dada JO (2017) An appraisal of paradigm shifts required of competence of the Nigerian quantity surveyors. Eng Constr Archit Manag 24(6):1269–1280

Dada JO, Jagboro GO (2012) Core skills requirement and competencies expected of quantity surveyors: perspectives from quantity surveyors, allied professionals and clients in Nigeria. Australas J Constr Econ Build 12(4):78–90

Elo S, Kyngas H (2007) The qualitative content analysis approach. J Adv Nurs Res Methodol 62:107–115. Available at: https://www.ncbi.nlm.nih.gov/pubmed/18352969. Accessed 20 Jan 2017

Garschagen M, Porter L, Satterthwaite D, Fraser A, Horne R, Nolan M, Solecki W, Friedman E, Dellas E, Schreiber F (2018) The new urban agenda: from vision to policy and action/will the new urban agenda have any positive influence on governments and international agencies? Informality in the new urban agenda: from the aspirational policies of integration to a politics of constructive engagement/growing up or growing despair? Prospects for multi-sector progression city sustainability under the NUA/approaching risk and hazards in the new urban agenda: a commentary/follow up and review of the new urban agenda. Plan Theory Pract 19(1):117–137. https://doi.org/10.1080/14649357.2018.1412678

Gu CC, Gomes T, Brizuela VS (2011) Technical and vocational education and training in support of strategic sustainable development. Thesis submitted for completion of Master of Strategic Leadership towards Sustainability, Blekinge Institute of Technology, Karlskrona, Sweden. Available at: https://www.diva-portal.org/smash/get/diva2:832727/FULLTEXT01.pdf

International Federation of Surveyors (2018) FIG guidance on professional competencies for quantity surveyors/construction economists/cost engineers. International Federation of Surveyors publication No. 17, Available at https://www.fig.net/resources/publications/figpub/pub71/Figpub71.pdf. Accessed 01 Jan 2021

International Labour Organisation (2011) A skilled workforce for strong, sustainable and balanced growth. A G20 training Strategy. International Labour Office—Geneva, 2010. Available at: www.ilo.org/skills/pubs/WCMS_151966/lang--en/index.htm?ssSourceSiteId. Accessed 20 Jan 2017

Krippendorff K (2004) Content analysis: an introduction to its methodology. Sage Publications, California

Kronner H (2005) The contribution of technical and vocational education and training to sustainable development. In: Proceedings of international workshop on workforce development for the knowledge economy, 7–13 Sept 2005, Seoul, Republic of Korea, 1–18

Leedy PD, Ormrod JE (2015) Practical research: planning and design, 11th edn. Pearson Education, New Jersey

Ma T, Luu HT (2013) The changing role of quantity surveyors in the green building development in South Australia. In: 38th AUBEA conference, University of Auckland, Australia

Mhlanga P (2018) Zimbabwe construction sector faces major hurdles. The Contractor 2018:1–36

Moyo T, Crafford G, Emuze F (2019) Decent working conditions for improved construction labour productivity on Zimbabwean building projects. Acta Structilia 26(2):1–38

Munzwa K, Wellington J (2010) Urban development in Zimbabwe: a human settlement perspective. Theor Empirical Res Urban Manag 5(14):120–146

Murray PE, Cotgrave AJ (2007) Sustainability literacy: the future paradigm for construction education? Struct Surv 25(1):7–23. https://doi.org/10.1108/02630800710740949

Naoum SG (2013) Dissertation research and writing for construction students, 3rd edn. Butterworth-Heinemann, Elsevier Ltd., Oxford

Omotayo T, Tan SW, Ekundayo D (2022) Sustainable construction and the versatility of the quantity surveying profession in Singapore. Smart Sustain Built Environ (Forthcoming). https://doi.org/10.1108/SASBE-07-2021-0125

Pacific Association of Quantity Surveyors (2001) Competency standard for quantity surveyors in the Asian-Pacific region. Available at: www.paquantitysurveyor.net. Accessed Dec 2 2016

Pappas E (2012) A new systems approach to sustainability: university responsibility for teaching sustainability contexts. J Sustain Educ 3:1–21 [Online] Available at: www.susted.org/. Accessed 2 June 2017

Perera BAKS, Hemajith SDM, Ginige K, Amaratunga RDG (2007) Quantity surveyor as the technical appraiser in the Sri Lankan financial industry. In: Built environment education conference University of Salford, Manchester, This version is available at: http://usir.salford.ac.uk/9827/

Piercy N, Rich N (2015) The relationship between lean operations and sustainable operations. Int J Oper Prod Manag 35(2):282–315. https://doi.org/10.1108/IJOPM-03-2014-0143

Presley A, Meade L (2010) Benchmarking for sustainability: an application to the sustainable construction industry. Benchmark: Int J 17(3):435–451. https://doi.org/10.1108/14635771011049380

Quantity Surveying Act Chapter 27:13 (2001) Zimbabwe Government Available at: www.law.co.zw/download/quantity-surveyors-act-chapter-2713/. Accessed 30 June 2022

Ramus J, Birchall S, Griffiths P (2008) Contract practice for surveyors, 4th edn. Butterworth-Heinemann, Oxford

Royal Institution of Chartered Surveyors (2018) The APC requirement and competencies. Royal Institution of Chartered Surveyors, London

Saunders M, Lewis P, Thornhill A (2016) Research methods for business students, 7th edn. Pearson Education, London

Seah E (2009) Sustainable construction and the impact of the quantity surveyor. In: Proceedings of the 13th Pacific Association of Quantity Surveyors (PAQS 2009), 36–48

Shayan S, Kim KP, Ma T, Fred R, Liu Z (2019) Emerging challenges and roles for quantity surveyors in the construction industry. Manag Rev: Int J 14(1):82–96

UN-HABITAT (2017) New urban agenda. UN-HABITAT, Nairobi

United Nations Economic Commission for Africa (2020) A harmonized implementation framework for the new urban agenda in Africa, publications and conference management section economic commission for Africa, Available at: https://unhabitat.org/a-harmonized-implementation-framework-for-the-new-urban-agenda-in-africa. Accessed 01 Jan 2021

United Nations Education Science and Cultural Organisation (2012) Shaping the education of tomorrow: 2012 report on the United Nations decade of education for Sustainable development, abridged. UNESCO digital library [Online] Available at: https://unesdoc.unesco.org

United Nations Education Science and Cultural Organisation (2014) UNESCO education strategy 2014–2021. UNESCO digital library [online] Available at: https://unesdoc.unesco.org

Valencia SC, Simon D, Croese S, Nordqvist J, Oloko M, Sharma T, Buck NT, Versace I (2019) Adapting the sustainable development goals and the new urban agenda: initial reflections from a comparative research project. Int J Urban Sustain Dev 11(1):4–23

Viertel E (2010) Vocational education for Sustainable development: an obligation for the European Training Foundation. Eur J Educ 45(2):217–235

Virakul B (2015) Global challenges, sustainable development, and their implications for organisational performance. Eur Bus Rev 27(4):430–446. https://doi.org/10.1108/EBR-02-2014-0018

Watson V (2016) Locating planning in the new urban agenda of the urban sustainable development goal. Plan Theory 15(4):435–448

World Economic Forum (2017) Harnessing public-private cooperation to deliver the new urban agenda. World Economic Forum

A Review of the Contribution of the Real Estate Sector Towards the Attainment of the New Urban Agenda in Zimbabwe

Ndarova Audrey Kwangwama, Shamiso Hazel Mafuku, and Yvonne Munanga

Abstract The real estate sector makes a significant contribution in an economy driving the shaping, configuration and transformation of the built environment to address the dynamic current and future needs of a society. This chapter reviews the contribution of real estate in Zimbabwe towards the attainment of the New Urban Agenda adopted in 2016 to expedite the implementation of Sustainable Development Goal 11. The New Urban Agenda provides a comprehensive framework to guide and track urbanisation globally underpinned on four core dimensions including social, economic, environmental and spatial sustainability. The real estate sector has played a significant role in the Zimbabwean economy from the colonial period through infrastructure projects, commercial, industrial and special purpose developments. However, from the year 2000 when government adopted the Fast Track Land Reform Programme, the real estate sector has experienced a myriad of challenges. The chapter reviews how the real estate sector has contributed towards the attainment of the NUA within a spatial planning framework that regulates its activities in the context of public interest to achieve sustainable development. The chapter is informed by the qualitative research approach. Data were collected through key informant interviews with officials from central government, local planning authorities and institutional property investors. The data were analysed through content review and thematic analysis. The study revealed that the outdated national land policy, spatial planning legislation and by-laws, poor land governance and administrative framework and the prevailing harsh economic environment have negatively affected the availability and servicing of land for urban development, significantly weakening the real estate sector. Therefore, the study recommends expediting conclusion of reviewing the national land policy, spatial planning legislation and the Model Building By-laws, setting up of a clear land governance and administrative framework and adoption of consistent favourable macro-economic policies to enable the real estate sector to meaningfully contribute towards attainment of the New Urban Agenda in Zimbabwe.

N. A. Kwangwama (✉) · S. H. Mafuku · Y. Munanga
Department of Architecture and Real Estate, University of Zimbabwe, Box MP 167, Mt. Pleasant, Harare, Zimbabwe
e-mail: akwangwama@gmail.com

C. Chavunduka and I. Chirisa (eds.), *New Urban Agenda in Zimbabwe*,
Advances in 21st Century Human Settlements,
https://doi.org/10.1007/978-981-97-3199-2_10

Keywords Urbanisation · Spatial planning · Real estate · New urban agenda · Sustainability

1 Introduction

The New Urban Agenda (NUA) of 2016 outlines the standards and principles that enable urban areas to be planned, constructed, developed, managed and improved on the basis of five core pillars that include national urban policies, urban legislation and regulations, urban planning and design, local economy and municipal finance and local implementation (UN-Habitat 2020, pp. x). The agenda incorporates a new acknowledgement of the interlinkages between sustainable urbanisation and development and it is closely linked to the 2030 Agenda for Sustainable Development, especially Goal 11 on sustainable cities and communities underpinned on social, economic, environmental and spatial sustainability (United Nations 2018a). Sustainable Goal 11 focuses on making cities and human settlements inclusive, safe, resilient and sustainable.

Construction and development in urban areas are driven by the real estate sector in response to demand for the different types of property. For example in 2014, globally, the value of real estate owned by institutional investors was estimated at U$32 trillion and the observed asset class allocations of institutional investors indicated that in Netherlands, Australia and Germany, real estate as a percentage of total assets held by institutional investors stood at 10%, 11% and 12%, respectively (Baum 2015). These figures illustrate real estate as a national key asset. Furthermore, national savings, investments and future pensions of any country are reflected through the real estate sector. Spatial planning provides the regulatory framework within which the real estate sector operates when executing real estate development projects shaping the built environment. It is in the context of analysing the role played by the real estate sector in the built environment of cities and towns that this chapter seeks to examine its contribution towards the attainment of the New Urban Agenda in Zimbabwe.

The real estate sector has played a significant role in the Zimbabwean economy from the colonial period through infrastructure projects, commercial, industrial and special purpose developments. The colonial government legislated an effective spatial framework that regulated and guided the evolution of a well-built environment. After the implementation of the Fast Track Land Reform Programme (FTLRP) by government in 2000 in Zimbabwe, the real estate sector started experiencing numerous challenges that include limited availability of land for urban development by local authorities, weakening economy, unavailability of financial resources for the development of both off-site and on-site infrastructure and environmental degradation and corruption. Compulsory acquisition of peri-urban farms negatively affected the availability of land with secure land rights while unavailability of finance impacted on the subsequent servicing of land for urban development. Most of the growth after the year 2000 became haphazard, unplanned and characterised by lack of services such as basic water and sanitation and lack of tenure security.

The growth of informal settlements in urban and peri-urban areas became a controversial issue. In 2018, estimates made by the World Bank indicated that 33.5% of the urban population in Zimbabwe are resident in informal settlements. In a study profiling informal settlements in Zimbabwe, Dialogue on Shelter profiled 60 slum settlements in Harare alone (Masimba 2021). In light of the various urban development challenges, this chapter is aimed at reviewing the contribution of the real estate sector in Zimbabwe towards the attainment of the New Urban Agenda of achieving spatial, social, economic and environmental sustainability in cities. The chapter therefore seeks to address the following questions:

(a) How does spatial planning influence the real estate sector in the shaping, configuration and transformation of the built environment of cities and towns in a sustainable way in response to needs arising from urbanisation?
(b) What has been the contribution of the real estate sector in Zimbabwe towards the attainment of the NUA in Zimbabwe?
(c) What challenges has the real estate sector experienced that have affected its effective contribution to the attainment of NUA in Zimbabwe?
(d) What policy measures should be adopted to enhance the contribution of the real estate sector towards the attainment of NUA in Zimbabwe?

2 Conceptualising Urbanisation, Spatial Planning, Real Estate and Sustainable Development

Urbanisation refers to the dynamic socio-economic process transforming human settlements from a rural to an urban form altering the geographical distribution pattern of population across the rural and urban territorial spaces (United Nations 2018b). The world has experienced rapid urbanisation in recent decades as evidenced by the 25% increase in 68 years between 1950 and 2018 when the urban's population stood at 30% and increased to 55% in 2050; and the world's urban population is estimated to be 68% (Ibid. 2018b). The Preliminary Report of the 2022 National Census in Zimbabwe has revealed that the country's share of urban population has increased from 33 to 39%—a 6% increase in the last decade (Zimstat 2022). In this context of rapid urbanisation, spatial planning becomes an important activity guiding and regulating the real estate sector in urban development.

Rapid urban growth without spatial planning has resulted in disastrous consequences that include rising poverty, ever-increasing housing demand, failure by local planning authorities to provide adequate housing and pre-requisite infrastructure services and subsequent sprouting of slums and squatter settlements (UN-Habitat 2009). Consequently, poorly managed and unplanned haphazard organic growth emerges in cities and towns as a serious threat to the sustainable development of human settlements. However, it has been acknowledged that urbanisation if properly managed, is a key driving factor that promotes economic growth, poverty reduction and attainment of human development (United Nations 2018b). The European and North American experiences of urbanisation during the nineteenth and twentieth centuries where urbanisation was accompanied by industrialisation and rapid economic growth have demonstrated that the process could result in sustainable development of prerequisite infrastructure such as roads, water, sewerage and electricity and provision of social services such as education and health facilities that are key to the attainment of sustainable development (UN-Habitat 2009).

The European Compendium of Spatial Planning Systems and Policies (1997) defines spatial planning as a public sector activity influencing the future of the spatial distribution of socio-economic activities with an aim of creating a well-co-ordinated territorial arrangement of land uses efficiently linked to each other and balancing the demands of development without compromising protection of the environment. The World Bank (2009) also defines spatial planning as a public sector activity designating and regulating land uses for the development of a community's physical and socio-economic well-being and efficient utilisation of land resources. Yeboah and Shaw (2013) define spatial planning as a state-driven activity focusing on the optimum allocation of land resources over competing land uses to attain the sustainable development of human settlements. Thus the main purpose of spatial planning is to create a healthy, safe and well-functioning built environment minimising negative socio-economic and environmental effects of private sector development.

Chigudu and Chirisa (2020) argue that the major focus of spatial planning is to deliver economic, social and environmental benefits that create a more stable and predictable environment conducive for investment and development. Such an environment is important for the achievement of national development. Gurran (2011) concurs with this observation highlighting that spatial planning provides a well-defined approach and policy framework for the effective co-ordination of key components of urban development that include housing, transport, energy, industry, public space, recreation and protection of biodiversity. As already highlighted in the introduction, spatial planning through formulation of national urban policies, urban legislation and regulations and urban planning and design plays a critical role in the attainment of the New Urban Agenda.

Real estate refers to land and the improvements attached to it including buildings (Baum 2015). The main components of real estate include land, agricultural estates, industrial, commercial and residential buildings (Harvey and Jowsey 2004). Enever et al. (2010) outline the key characteristics distinguishing real estate from other investment opportunities that include its heterogeneity, indivisibility, inelastic supply, high transfer costs, special management challenges, special and unique risks, perpetual existence, imperfect knowledge and high government intervention. As a field of study, real estate encompasses valuation, real estate management, facilities management and real estate development. However, this chapter focuses on real estate development that is regulated by spatial planning in the creation of the built environment of cities and towns in response to needs emanating from urbanisation. The chapter thus analyses the contribution of real estate sector in Zimbabwe through real estate development towards the attainment of the New Urban Agenda.

Real estate developers are key stakeholders in the built environment as they undertake different forms of development in response to demand for different types of real estate products. Any society experiences dynamic changes on its demographic structure, incomes, tastes, level of industrial and technological development and advancements (Miles et al. 2007). In response to such dynamic changes largely emanating from urbanisation, developers undertake different forms of real estate development that include new developments, conversion of existing uses, refurbishment of existing buildings and redevelopment (Harvey and Jowsey 2004). During the implementation of the different forms of real estate development, developers are concerned with maximisation of profit. However, spatial planning as a public sector activity, intervenes in the built environment to regulate real estate development in the public interest.

Adams (1994) highlights that spatial planning as a public sector activity intervenes in the built environment through preparation of statutory plans, development control and development promotion. Statutory plans provide a guiding and regulating framework to real estate developers in respect of where a local planning authority is encouraging development, preventing it and directing it. Local planning authorities use development control as an administrative machinery to direct particular development proposals and make decisions on a case by case basis upholding provisions of the statutory plan or diverging from it. In development promotion, spatial planning focuses on promoting and marketing locations and making land available to developers and providing grants and subsidies for depressed regions that are less attractive (Adams 1994). In concurrence with this observation, the European Compendium of Spatial Planning Systems and Policies (1997) argues that spatial planning encompasses national and transnational planning, regional policy, regional planning and detailed land use planning. Thus spatial planning entails co-ordination of sectoral policies and economic development to ensure their fair distribution across territorial space and regulation of land and building uses that would not otherwise have been implemented if only driven by market forces.

Sustainable development is defined as development that meets the needs of the present without compromising the ability of future generations to meet their own needs (The Brundtland Commission 1987: 43). Adams (1994) highlights that in spatial planning, the principle of sustainable development questions the entrenched power of short-term costs and benefits of a development in preference of longer-term costs and benefits brought in by longer-term visions. Miles et al. (2007) indicate that that sustainable environmental quality is achieved through the various spatial planning measures that include using building designs which capture the benefits of renewable resources such as solar and geothermal energy, site designs that sustain critical natural landscape and hydrological features bordering and within developed areas, promoting the recycling of manufactured materials, use of natural lighting and ventilation systems and on-site storm water infiltration.

3 Theoretical Framework and Literature Review

3.1 Theoretical Framework

As already highlighted in the introduction, the NUA outlines the standards and principles in respect of how urban areas should be planned, constructed, developed, managed and improved on the basis of national urban policies, urban legislation and regulations, urban planning and design, local economy, municipal finance and local implementation. The focus of this chapter is to assess the contribution of the real estate sector towards the attainment of the NUA in Zimbabwe. This section reviews contemporary spatial planning theories, namely new urbanism and smart growth that originated in the United States of America but have provided the planning framework guiding contemporary urban development across the globe. The real estate sector drives the implementation of new urbanism and smart growth ideas and visions as it undertakes various projects including commercial, industrial and residential developments.

New urbanism originated as an architectural design-oriented movement that focuses on design, scale, mixed land uses, improved street network, promotion of social cohesion and sense of place (Berke et al. 2006). Members of the New Urbanism Movement adopted a Charter in 1996 that outlined twenty-seven principles for the scales of development ranging from the region, city to the town. The key design principles of the new urbanism include walkability, discouraging increased use of the automobile, connectivity, mixed use and diversity, compact buildings, discouraging leap frog developments that foster urban sprawl, increased density, preservation of open spaces, promoting community identity and high quality of life (Wey and Hsu 2014). Thus, the new urbanism argues for the creation of dense, compact and mixed-use urban developments that promote diverse closely-knit communities with pedestrian-friendly streets.

Smart growth advocates for the designing of compact mixed-use developments, enhanced mobility through provision of various modes of transportation, improved livability of created places where people live, work and play and designing infrastructure systems that are efficiently managed (Berke et al. 2006; Miles et al. 2007). Duany et al. (2009) highlight that smart growth shares common visions and attributes with the new urbanism fostering mixed land uses, compact developments, creation of walkable neighbourhoods, provision of wide transportation modes, preservation of open spaces, agricultural land, areas of scenic beauty and environmentally sensitive areas and promotion of distinctive and attractive communities with a strong sense of place. Furthermore, citizens and stakeholders are encouraged to participate in making development decisions. Such decisions should be predictable, fair and effective in cost management. In terms of housing provision, smart growth encourages the creation of affordable housing schemes providing a wide range of options and choices for different household types, family sizes and income levels (Berke et al. 2006). Thus smart growth focuses on macro-scale land use and policy framework centred on spatial planning and public policy principles.

Although urbanism and smart growth share common visions and attributes, they have differences. The new urbanism offers design principles that are more detailed and site-specific compared to smart growth that focuses more on spatial planning and public policy principles as already pointed out earlier on. In comparison with smart growth, the new urbanism also provides less guidance and influence in protecting environmentally sensitive sites, regeneration of degenerating inner cities and delivery of affordable houses (Berke et al. 2006). Furthermore, since new urbanism evolved as an architecturally oriented approach, its main focus is on physical design images and neighbourhood layouts. Despite their minor differences, the two theories have greatly influenced contemporary spatial planning in the development of human settlements.

3.2 The New Urban Agenda and Real Estate Development

As highlighted earlier on, the NUA was adopted in October 2016 at the United Nations Conference on housing and sustainable development. The agenda advocates for a better and more sustainable future where all people have equal rights and equal access to the benefits that cities offer, as well as the opportunities they present. The NUA endeavours to promote the achievement of the 2030 Sustainable Development Goals. It appreciates and acknowledges that the current era of rapid urbanisation if not properly managed in most developing countries may present challenges to urban communities. On the other hand, if well-planned and properly managed, this urbanisation can be a powerful tool for sustainable development (United Nations 2018b). While SDG 11's objective is making cities and human settlements inclusive, safe, resilient and sustainable, the New Urban Agenda complements it as it lays out the standards and principles for planning, construction, development, management and improvement of urban settlements. Therefore, the NUA calls for a transformation or reconsideration of the urban systems and physical form of urban spaces so as to achieve SDG 11.

The interlinkages of SDG 11 with the NUA are extensive and wide ranging. Both have a direct connection to urban policies which have a clear impact on cities and human settlements (United Nations 2018a). These cities and human settlements are delivered through real estate which is governed by these urban policies to ensure development of cities that are safe, inclusive, sustainable and resilient. The real estate development process undertakes diverse projects ranging from residential, commercial, industrial and recreational including the requisite physical infrastructure. Neuman and Smith (2010) assert that great cities are born of and give rise to great infrastructure and in history, city planning has been intensely engrained in real estate development and physical planning.

Real estate development has the potential to advance sustainability in terms of fulfilling physical, social and economic development goals of any nation such as providing local employment and creating economic growth through efficient utilisation of resources, both human and material to construct new or renovate existing buildings and services. Furthermore, real estate development acknowledges that growth in real estate is a business of multiple aspects encompassing operations from the renewal of existing buildings and re-rentals to the purchases of undeveloped land and sales of developed lots (Kaklauskas et al. 2015). Land is at the core of NUA as land governance underpins all key aspects of the agenda. Access and use of land coupled with good land governance are critical if a nation is to harness the transformative potential of the NUA. This implies that governments should put measures in place to ensure affordable and equitable access to land and tenure security for all. Such measures can be promoted through establishment of transparent, participatory and accountable frameworks that ensure the sustainable use of land. Furthermore, governments should put in place mechanisms to generate revenue from land (Satterthwaite 2016).

NUA intends to reduce inequalities that exist in cities, promote inclusive and sustainable economic growth, improve human health and well-being, foster resilience and protect the environment by re-addressing the way cities and human settlements are planned, designed, financed, developed, governed and managed. The New Urban Agenda also envisages cities that fulfil their social functions regarding adequate provision of housing and prerequisite infrastructure (United Nations 2018a). Real estate development plays a key role in ensuring attainment of the key objectives of NUA through implementation of various forms of projects that are compliant with these objectives.

To ensure an optimal mix of development that complies with the current spatial planning frameworks and at the same time ensuring developments that are environmentally, socially, spatially sustainable and economically efficient, the real estate sector across the globe has embraced the mixed-use development concept which has taken centre stage in urban planning and real estate development the world over in the past few decades (Herndon 2011). The mixed land use principle is a ubiquitous component of the underlying visions and ideals of contemporary spatial planning theories and concepts such as new urbanism, smart growth and the compact city (Rabianski et al. 2009; Neuman and Smith 2010; Herndon 2011). Current spatial planning frameworks have been created on the basis of these contemporary theories. The real estate sector undertakes all its diverse real estate projects within such spatial planning frameworks complying with all the principles.

There is a general consensus in the literature that real estate development is a complex process (Healey and Barret 1990; Gore and Nicholson 1991; Wilkinson and Reed 2007; Drane 2013). Several models have been developed to understand the real estate development process that includes the equilibrium, the event sequence, the agency, the structural model and the institutional models (Healey and Barret 1990; Healey 1991, 1992). It has however been acknowledged that there is no single model appropriate for all circumstances.

The equilibrium model has evolved from neo-classical economics. The model assumes that effective demand drives real estate development. Different players in the real estate industry respond to market signals implementing real estate projects to satisfy demand in a way Linchfield and Drabkin (1980) describe as at the right time, place and price. Healey (1991) highlights that the market signals indicating demand are reflected in the rental and yield levels. Developers exploit this information and initiate real estate projects that supply new property. However, there are several criticisms against the equilibrium model. Healey and Barret (1990) argue that the model does not take into account the different types of demands that respond to different signals. Moreover, real estate development is not only driven by demand but by other factors that are not necessarily economic. The development process is also complex and some of the valuation approaches used in assessing the value of properties create distortions when deriving yields.

The event sequence model views the real estate development process as a multi-faceted activity with a series of stages involving various players. The model is based on the description of the key tasks involved in the development process by focusing primarily on tracing certain perceived key activities that occur simultaneously within each project. In that regard, the development process is perceived as an eight-stage process described as initiation, evaluation, acquisition, and design and costing, planning permissions, commitment and lastly implementation and disposal (Wilkinson and Reed 2007). The model has however been criticised as being too descriptive and lacks specification of the actors involved and how their roles, interests and strategies affect the entire development process (Healey 1991).

Despite the criticism against the event sequence development model, it is still relevant in upholding the standards and principles for planning, construction, development, management and improvement of urban settlements which is the main focus of the New Urban Agenda. However, continuous sprawl and peri-urban growth in developing countries has resulted in the sprouting of settlements without layout plan approvals and no basic infrastructure and various stages in the event sequence model have been disregarded in the development of various peri-urban settlements. The agency model focuses on the various players in the real estate development process, their roles and the interests guiding their strategies (Healey 1991). The model describes the relationships among the players from a behavioural and institutional perspective. However, the main weakness of this model is that it does not explain the driving forces of the real estate development process.

The structure model is based on the urban political economy theories, i.e. theories of the structure and the dynamics in the production and exchange of commodities. The model seeks to explain the structure of markets through power relationships of capital, labour and owners of land. In the context of property development, the structure model focuses on how real estate development is structured by relationships between capital and labour, capital and owners of land and between state regulation and the market (Healey 1991). The main weakness of the model is that it does not explain the real estate development process events and the connections between relationships of different players.

The institutional model propounded by Healey (1992) encompasses complex events, various agencies engaging in the real estate development process and the various forms through which the development process has manifested under distinctive forms. Healey (1992) identifies four stages of the institutional model that includes describing events constituting the development process and the agencies initiating them, identifying the roles and how they are related in respect of power, assessing the strategic decisions and actions influencing the roles and the manner government regulations and policies influence these roles and the relationships between them. Since the institutional model is a consolidation of the other four development models (equilibrium, event sequence, agency and structural), it is more realistic at addressing the weaknesses of these models.

4 A Brief Historical Synopsis of Real Estate Development in Zimbabwe

In order to understand the role of the real estate sector in the evolution of the built environment in Zimbabwe, it is critical to undertake a historical synopsis of development of the built environment from the colonial occupation of the country in 1890. The evolution of the built environment is closely linked to the urbanisation process and the development of the colonial economy. The early period of colonisation (1890–1945) laid down the foundation of real estate development in Zimbabwe through land alienation from the indigenous people, privatisation and creation of the colonial spatial structure (Wekwete 1989; Zinyama 1993). The alienation and privatisation of land were the key underlying cornerstones of the colonial economy that created the base for the establishment of new urban settlements anchored on a real estate sector that implemented residential, commercial and industrial developments.

Real estate development in the new colony was characterised by annexation and subdivision of agricultural land adjoining urban centres. For example in Harare, the first major subdivision of agricultural land was undertaken in 1920 when Avondale farm located six kilometres from the Central Business District (CBD) was subdivided and disposed to property developers for residential development and suburban commercial uses (Wekwete 1989; Zinyama 1993). The first Town Planning Act was legislated in 1933 against a background of significant urban development in the new colony and as part of consolidating the development framework of the emerging built environment. This legislation empowered municipalities to prepare urban planning development schemes (Wekwete 1987).

In 1945, the colonial government repealed the 1933 planning legislation and enacted another Town Planning Act in 1945 that consolidated urban planning schemes, empowered local authorities to compulsorily acquire land for future urban expansion and regulated evolving townships and control land subdivisions (Wekwete 1987). The schemes focused on the preparation of land and building regulations and provided a strong base for an effective development control system later reinforced by the enactment of the Regional, Town and Country Planning (RTCP) Act (Chapter 29:12) in 1976. The schemes established a high degree of certainty in the land and real estate market required by real estate owners and investors.

Furthermore, the colonial state's industrialisation policy stimulated the growth of the land and real estate market. Large tracts of peri-urban agricultural land were subdivided into residential stands. For example in Salisbury (now Harare), 1228 residential stands in Mt. Pleasant were subdivided over a ten-year period between 1940 and 1950 (Zinyama 1993; Wekwete 1994). More large-scale residential subdivisions were prepared and approved creating most of the low-density suburbs in Harare such as Marlborough and Greendale (Zinyama 1993; Wekwete 1989). Residential real estate development was further boosted by the enactment of the Building Societies Act in 1951. This legislation created the much-needed financial framework for European residential suburban developments complementing the post-war housing fund used as an incentive to attract European immigrants into the country.

The mortgages advanced by building societies only benefited Europeans. Africans lived in rented accommodation provided by the municipalities. The Urban Councils Act of 1973 required each urban local authority to operate an African Affairs Account which would generate revenues from the sale of traditional African beer enterprises to finance most capital and recurrent expenditure. Musekiwa (1993: 55) has argued that this was the most curious paradox of colonial history that state provided social services using finance raised from the sale of traditional beer to Africans who were the recipients of the service.

In 1965, Southern Rhodesia (now Zimbabwe) made a Unilateral Declaration of Independence (UDI) from Britain resulting in international sanctions being imposed on the country. In response, the colonial government adopted a vigorous import substitution policy strengthening local backward and forward linkages. More employment was created in urban areas resulting in increasing demand for office, industrial and residential accommodation with the highest demand for residential stands experienced during the period 1965–1971 for both Europeans and Africans. The built environment was further consolidated with the creation of new industries and the state's intervention in infrastructure development. In order to effectively control and manage physical development of land and buildings in the built environment, the colonial government enacted the Regional, Town and Country Planning Act (RTCP) Act in 1976 replacing the 1945 Act. This Act formed the centre piece of town planning legislation and was complemented by other legislations such as Housing and Building Act and the Model Building By-Laws in the regulation of the built environment.

The first two decades of independence, i.e. 1980–2000, were characterised by the removal of racial discrimination and removal of laws that had restricted Africans from migrating into urban areas resulting in high demand for residential accommodation and ancillary infrastructure services (Musekiwa 1993; Zinyama 1993). The new government adopted the home ownership policy which encouraged Africans to acquire their rented accommodation from municipalities. More residential stands were also availed through self-help and the sites and services schemes. Finance for these residential schemes was availed through the Public Sector Investment Programmes and the National Housing Fund (Chenga 1993; Musekiwa 1993). Municipalities also accessed loans from multilateral institutions such as the World Bank to finance the development of both off-site and on-site infrastructure. The new government also adopted the growth point policy which established new rural growth points providing infrastructure to the formerly deprived communal areas. The national economy was fairly stable during the first twenty years of independence with the country accessing loans from multilateral institutions.

As highlighted earlier on, in 2000, the Government of Zimbabwe implemented the FTLRP. One of the major consequences of the FTLRP was that Zimbabwe experienced significantly continued economic decline from 2000 to 2008 often described as the 'lost decade' (African Development Bank 2011). During this period, the national economy was characterised by acute shortages of foreign currency, severe shortages of basic commodities due to poor agricultural performance and low capacity utilisation by the manufacturing sector, de-industrialisation, continued decline of the country's gross domestic product, intermittent supply of public utilities resulting in an ailing public health system, severe price distortions due to multiple foreign exchange markets, serious brain drain as skilled people migrated out of the country seeking greener pastures in the region and overseas, decline of foreign direct investment and high levels of corruption. These problems have affected the effectiveness of the construction industry, a key sector in the built environment.

The continuously deteriorating macro-economy also negatively affected the availability of financial resources to local authorities to upgrade old infrastructure and development of new infrastructure (Dohwe and Kwangwama 2019). For example, all the five sewage treatment plants in Harare are currently overloaded requiring urgent upgrading. The overloaded treatment works have partially treated the sewage that is discharged into the dams supplying water to Harare compromising environmental quality from water pollution. Furthermore, the capacity of the existing dams that supply water to the Harare Metropolitan Region has been exceeded and Kunzvi Dam has been on the drawing board since the 1990s. Thus major low and medium residential projects cannot be connected to the existing water and sewage reticulation.

5 Research Methodology

This chapter is informed by an extensive literature review and the qualitative research approach. Data collection was conducted through key informant interviews. A total of ten key informant interviews were conducted with institutional property investors, real estate professionals, officials from both local planning authorities and central government departments. The data were analysed through content review and thematic analysis.

6 Research Findings

This section presents the research findings on how the real estate sector is contributing towards the attainment of the New Urban Agenda in Zimbabwe. Key informant interviews with officials from both local planning authorities and central government revealed that undertaking any real estate development projects requires land for the development of buildings and infrastructure. Land for real estate development in Zimbabwe is availed by land owners that include the state, local authorities and the private sector. Real estate development is undertaken by developers from the private sector, local authorities and central government. Private developers are the key players in commercial and real estate projects while local authorities and central government are actively involved in public sector housing development. All the different types of real estate require land for development.

Officials from local planning authorities highlighted that land availability in large cities such as Harare became a very controversial issue after the year 2000 which marked the beginning of the FTLRP. Developers are currently experiencing problems in accessing land to undertake their real estate development projects. The land allocation system has been characterised by corruption. In one of the key informant interviews with institutional property investors, one of the key informants revealed how they approached an indigenous company as a 'front' to obtain land from the local planning authority to develop an office park. The land was portion of a golf course leased by the local planning authority to a private golf club. The indigenous company approached the local planning authority which repossessed portion of the golf course. A change of reservation application was made to the Department of Physical Planning which issued a permit granting the change of reservation from recreational use to offices. The processing of the sale of the subject land was conducted in the name of the indigenous company yet the actual purchaser of the land was the institutional property investor. If the institutional investor had not used the indigenous company as a 'front' to access the land, they would not have obtained it.

Key informant interviews with officials from local planning authorities revealed that prior to the FTLRP in 2000, both local planning authorities and private real estate developers could negotiate to acquire land directly with owners of farms located in peri-urban zones for new real estate development projects. They would negotiate the acquisition of the land which would be zoned for agricultural uses. The transactions would be conducted on a willing buyer willing seller basis. Consequently, local authorities accumulated large land banks for real estate development which they would use themselves or allocate to private developers for housing, commercial and industrial developments. Private developers could also acquire land directly from land owners, a typical example being the development of Westgate Regional Shopping Centre and residential area in Harare. Old Mutual Property Zimbabwe (Private) Limited acquired the land which was a farm on the peri-urban area of Harare for real estate development, prepared a local development plan and developed a regional shopping centre and serviced stands for medium density housing and other ancillary uses.

One of the consequences of the FTLRP was the gazetting of most of the farms located in the peri-urban areas of large cities for compulsory acquisition for rural resettlement. Key informant interviews with both central government and local planning authorities revealed that administratively, all such gazetted farms now fall under the Ministry of Lands, Agriculture, Fisheries, Water and Rural Resettlement. Procedurally, if the gazetted peri-urban land is required for urban development, the Minister of Local Government and Public Works will approach the Minister of Lands, Agriculture, Fisheries, Water and Rural Resettlement who will release the acquired land. The Minister of Local Government and Public Works will subsequently allocate the released land to the relevant local planning authority which should in turn avail it to private developers for real estate development. Key informant interviews with both local planning authorities and institutional property developers highlighted that this process is tedious, too bureaucratic and politically motivated. The interviews also highlighted that land barons and politically aligned housing co-operatives took advantage of the FTLRP and invaded a number of peri-urban farms adjoining cities. In the case of Harare, haphazard, chaotic informal housing developments have emerged in Caledonia, Harare South, Harare North and Snake Park. These areas do not have proper planning and adequate provision of both on-site and off-site infrastructure compromising the four pillars of sustainability under the New Urban Agenda: spatial, social, economic and environmental sustainability.

Key informant interviews with officials of both central government and local planning authorities also revealed that despite the dynamic changes brought by the FTLRP, land governance in Zimbabwe is still guided by the 1990 Zimbabwe Land Policy. The provisions in this document are archaic and no longer relevant to address the emerging challenges. To address challenges arising from this existing outdated land policy, the Ministry of Lands, Agriculture, Fisheries, Water and Rural Resettlement commenced the review process of the existing Zimbabwe Land Policy Document in 2020. However, at the time of writing this chapter, the review process had not yet been completed.

The unavailability of land for real estate development has also resulted in some developers encroaching on wetlands. Key informant interviews with officials from both central government and local planning authorities revealed that there have been legal battles between the Environmental Management Agency (EMA), the public and developers. EMA argues that wetlands should not be developed but left natural as they are. They provide society with fresh water and provide a bio-diversity environment for amphibians, reptiles as well as bird breeding and migration. They are very useful in mitigating climate change as vegetation in these wetlands acts as a carbon reservoir reducing the amount of carbon dioxide in the atmosphere and decrease the greenhouse effect. The construction of houses on wetlands has compromised the fourth pillar of the New Urban Agenda, that is, environmental sustainability. In 2020, several local planning authorities invoked demolition orders on houses built on wetlands. Consequently, several residential properties were demolished in Chitungwiza, Budiriro in Harare and other towns throughout the country.

Key informant interviews with officials from central government, local planning authorities and the institutional property developers revealed that Zimbabwe inherited a well-regulated built environment from the colonial period with a spatial planning legislative framework guiding and regulating real estate development. The RTCP Act still forms the centrepiece of planning legislation complemented by other legislations in regulating physical development in the built environment. No significant changes have been made to the RTCP since its promulgation in 1976 but there have been fundamental changes arising from rapid urbanisation and population growth after independence.

Key informant interviews with institutional property developers highlighted some of the archaic provisions such as the *deemed refused* response by local planning authorities (LPA). This provision states that if an LPA does not determine a development or subdivision application within four months of the date the application was acknowledged, it is deemed refused. The institutional developers argued that such archaic provisions required amendment as they encourage complacency by LPA officials but disadvantaging developers. Furthermore, some of the statutory plans such as town planning schemes prepared during the 1970s are still operative despite the significant social, economic and physical changes that have occurred over the years. A significant number of urban local authorities are also operating without statutory plans such as master and local development plans to guide development making it difficult for town planners to regulate the built environment.

The Model Building By-Laws were formulated in 1977 and no significant amendments were made since they were published. Key informant interviews with officials from both central government and local planning authorities acknowledged that some of the provisions in the RTCP Act, old town planning schemes and other regulations and by-laws are outdated negatively affecting the efficient processing of town planning development and subdivision permits and regulation of developments resulting in the failure to attain spatial sustainability, one of the key pillars of the New Urban Agenda.

Key informant interviews with officials from central government revealed that the Department of Spatial Planning and Development has also initiated the comprehensive review of the various town planning principles enshrined in the existing RTCP Act on the preparation of the different statutory plans, development control, regularisation of illegal settlements, subdivisions and consolidations. The key informant interviews also revealed that local authorities such as City of Harare are reviewing some of the existing outdated town planning schemes and replacing them with local development plans. For instance Avondale Local Development Plan 39 and Arundel Local Development Plan 53 replaced the outdated town planning schemes for their areas. In 2021, the Ministry of National Housing and Social Amenities also commenced the public engagement process inviting suggestions to amend the existing 1977 Model Building By-Laws.

Key informant interviews with institutional property developers revealed that the successful contribution of real estate development to the built environment heavily depends on the availability of finance. Private sector real estate developments have been largely financed from financial institutions including building societies and insurance and pension funds. The commercial and industrial sectors were largely driven by insurance and pension funds who own most of the buildings in the Central Business Districts (CBDs) of most cities and towns in Zimbabwe. However, since the continued deterioration of the national economy after the year 2000, most of these institutions have focused investment of their funds on the residential sector for example Fidelity Life Insurance developed Manresa and South View Park residential schemes while Budiriro housing project was jointly undertaken by Old Mutual and Central Africa Building Society (CABS), all in Harare.

Key informant interviews with local planning authorities however highlighted that the previously mentioned residential schemes have been out of reach of most low-income families currently engaged in the informal sector. Such low-income families have fallen prey to housing co-operatives and land barons who have been allocating unserviced residential areas located on the edges of most urban areas. As already highlighted, such informal settlements do not have both off-site and on-site infrastructure with residents relying on boreholes and shallow wells for drinking water and using ventilated improved pit (VIP) latrines. These informal settlements are breeding grounds for waterborne diseases such as cholera and typhoid compromising the attainment of social sustainability in the New Urban Agenda.

Key informant interviews with institutional property developers highlighted that after 2015, real estate developers focused on densification, redevelopment and conversions in residential areas. Densification entails development of a site to its maximum potential. For example, developers in Harare have been constructing cluster houses on low-density residential sites instead of building single detached residential properties. Developers have also focused on demolishing old dilapidated buildings in residential areas adjoining the CBD and redeveloping new properties. For example in the Belgravia area of Harare, several old residential properties have been pulled down and replaced by new office developments. Furthermore, the same area has experienced conversion of properties from residential to office use. These conversions have largely been driven by new demand for suburban offices due to problems in the Central Business District such as high building operating costs, traffic congestion and street kids. Although there has been recognition on the need for urban renewal in areas such as the Harare Kopje, Mbare Siyaso and Mbare Hostels and Sakubva in Mutare, lack of political will and inadequate financial resources have affected the successful implementation of the urban renewal projects.

7 Discussion, Conclusion and Recommendations

During the colonial and post-colonial periods, but prior to 2000, real estate development was undertaken through properly laid down procedures of statutory planning, land banking and estate development with the formal acquisition of peri-urban agricultural land and its subsequent incorporation into towns and cities for development into the built environment. Subsequent to the acquisition, the land was carefully planned, zoned and serviced with requisite infrastructure for real estate development using funds from estate development accounts of local authorities. Real estate development during this period complied with the event sequence model and Healey's institutional model where development followed certain stages and the state provided the institutional and legal framework regulating the real estate sector.

However, the year 2000 marked the beginning of the FTLRP which designated all peri-urban farms for acquisition which has negatively affected land availability for real estate development. Non-availability of land is also linked to other myriad of problems including invasion of wetlands, continued deteriorating macroeconomic environment, lack of adequate financial resources, obsolete infrastructure and outdated development framework. The existing spatial planning legislation and by-laws are outdated and delay the processing of town planning required in real estate development. The implications of such delays have negatively affected the successful contribution of the real estate sector on the realisation of the New Urban Agenda specifically the four pillars that include spatial, environmental, social and economic sustainability.

This chapter recommends that central government expedites the conclusion of reviewing the national land policy, spatial planning legislation and Model Building By-laws, setting up of a clear land governance and administrative framework and adoption of consistent favourable macro-economic policies to enable the real estate sector to meaningfully steer and drive the New Urban Agenda in Zimbabwe. The reviewed national land policy and land governance and administrative framework will avail land required for real estate development while the reviewed spatial planning legislation will promote efficient processing of development permits required by the real estate sector prior to commencement of any developments. The reviewed Model Building By-Laws will improve the regulation of building construction in the built environment. Addressing the macro-economic environment will create a conducive environment for financing the upgrading of old infrastructure and the development of new infrastructure. All these anticipated changes in regulating the built environment will positively influence the attainment of the New Urban Agenda.

References

Adams D (1994) Urban planning and the development process. UCL Press, London

African Development Bank (2011) Infrastructure and growth in Zimbabwe: an action plan for sustained strong economic growth. Tunis, Tunisia

Baum A (2015) Real estate investment: a strategic approach. Routledge, London

Berke PR, Godschalk DR, Kaiser EJ, Rodriguez DA (2006) Urban land use planning. University of Illinois Press, Chicago

Brundtland GH (1987) Report of the World Commission on environment and development: "Our common future." United Nations, New York

Chenga MM (1993) Provision of housing in Zimbabwe. In: Zinyama LM, Tevera DS, Cumming SD (eds) Harare: the growth and problems of the city. University of Zimbabwe Publications, Harare, pp 45–50

Chigudu A, Chirisa I (2020) The quest for a sustainable spatial planning framework in Zimbabwe and Zambia. Land Use Policy 92:10442

Dohwe P, Kwangwama NA (2019) Urban resilience in cities of the developing world with reference to Harare, Zimbabwe. Chapter 6 In: Chirisa I, Mabeza C (eds) Community resilience under the impact of urbanisation and climate change: cases and experiences from Zimbabwe. Research and Publishing CIG, Bamenda, Cameroon, pp 53–94

Drane J (2013) The state of contemporary property development theory. In: 19th Annual Pacific-Rim real estate conference, Melbourne, Australia, pp 13–16

Duany A, Speck J, Lydon M (2009) Smart growth manual: new urbanism in American communities. McGraw Hill, New York

Enever N, Isaac D, Daley M (2010) The valuation of property investments. Estate Gazette, London

European Commission. Directorate for Regional Policy and Cohesion (1997) Compendium of European Planning Systems. Regional Development Studies, vol 28. Office for Official Publications of the European Communities, Luxembourg

Gore T, Nicholson D (1991) Models of the land development process: a critical review. Environ Plan A 23(5):705–730
Gurran N (2011) Australian urban land use planning: principles, systems and practice. Sydney University Press
Harvey J, Jowsey (2004) Urban land economics. Macmillan Press, London
Healey P (1991) Models of the development process: a review. J Prop Res 8(3):219–238
Healey P (1992) An institutional model of the development process. J Prop Res 9(1):33–44
Healey P, Barret S (1990) Structure and agency in land and property development processes: some ideas for research. Urban Studies 27(1):89–104
Herndon JD (2011) Mixed-use development in theory and practice: learning from Atlanta's mixed experiences. University of Georgia, Georgia
Kaklauskas A, Zavadskas EK, Dargis R, Bardaus K (2015) Sustainable development of real estate. VilnusGediminas Technical University, Vilnus
Linchfield N, Drabkin D (1980) Land policy in planning. George Allen and Unwin, London
Masimba G (2021) Harare city scoping study. Harare, Dialogue on Shelter Trust. Available Online: https://www.africancities.org/wpcontent/uploads/2021/12/ACRC_Harare_City-Scoping-Study.pdf. Accessed 18 Sept 2022
Miles ME, Berens GL, Eppli MP, Weiss MA (2007) Real estate development: principles and process. Urban Land Institute, U.S.A
Musekiwa A (1993) Low-Income Housing Development in Harare: A Historical Perspective. In: Zinyama LM, Tevera DS, Cumming SD (eds) Harare: The Growth and Problems of the City. University of Zimbabwe Publications, Harare, pp 51–60
Neuman M, Smith S (2010) City planning and infrastructure: once and future partners. Sage, Texas
Rabianski J, Gilber K, Clements J (2009) Mixed-use development: a call for research. ULI, Washington DC
Satterthwaite D (2016) A new urban agenda. Sage Publications, New York
UN-Habitat (2009) Planning sustainable cities: global report on human settlements. Earthscan, United Kingdom
UN-Habitat (2020) The new urban agenda illustrated. UN-Habitat, Nairobi, Kenya
United Nations (2018b) World urbanisation prospects: the 2018 revision. United Nations, Department of Economic and Social Affairs, New York
United Nations (2018a) High-level political forum on sustainable development: 2018 review of SDGs implementation: SDG 11—make cities and human settlements inclusive, safe, resilient and sustainable, Habitat II, New York
Wekwete KH (1989) Physical planning in Zimbabwe: a review of the legislative, administrative and operational framework. Town Plan Rev 11(1):49–69
Wekwete K (1987) Development of urban planning in Zimbabwe—an overview. RUP Occasional Paper No. 8. Harare, Department of Rural and Urban Planning, University of Zimbabwe
Wekwete K (1994) Urbanisation, urban development and management in Zimbabwe. In: Wekwete KH, Rambanapasi CO (eds) Planning urban economies in Southern and Eastern Africa. England, Avebury, pp 9–54
Wey WM, Hsu J (2014) New urbanism and smart growth: toward achieving a Smart National Taipei University District. Habitat Int 42:164–174
Wilkinson, Reed R (2007) Property development. Routledge, London
World Bank (2009) Doing business 2010. World Bank, Washington, DC
Yeboah E, Shaw DP (2013) Customary land tenure practices in Ghana: examining the relationship with land-use planning delivery. Int Dev Plan Rev 35(1):22–39
ZimStat (2022) 2022 Population and housing census: preliminary report on population figures. ZimStat, Harare
Zinyama, L. (1993) The evolution of the spatial structure of Greater Harare: 1890 to 1990. In: Zinyama LM, Tevera DS Cumming SD (eds) Harare: the growth and problems of the city. University of Zimbabwe Publications, Harare, pp 7–32

Institutions, Laws and Governance Structures for Developing and Managing the Built Environment: Elephant in the Room for Advancing the New Urban Agenda

Simbarashe Show Mazongonda and Andrew Chigudu

Abstract The New Urban Agenda (NUA) seeks to ensure a better and inclusive future by balancing out economic, social and environmental sustainability. The agenda acknowledges the complexity of urban environments as myriad land uses; housing, industrial, recreational, institutional, and commercial. Such a mixture of land uses poses sustainability challenges. One of the pillars of the NUA speaks to the importance of urban legislation and regulations in fulfilling, protecting and respecting private and public interests. Urban development takes place in the context of procedures clearly outlined in policies, statutory plans, acts of parliament and statutory instruments. Some sections of literature have labelled current legislation and policies as outdated and not keeping pace with changing realities. If the context, policy and institutional framework for achieving sustainable cities in Zimbabwe is not properly structured, analysed and understood, smooth urban management systems will remain a dream in the pipeline. This chapter provides a prescriptive analysis of institutions and laws that impinge on the NUA with the view of exposing policy and practical bearings by different stakeholders. It largely relies on secondary data. Documentary analysis informed by analytical rationality was used to skim, collate, and reorder data from a sizeable number of sources. Examples drawn from urban informality, urban economics, technological factors, environmental conservation, planning process and governance quality were used to show how a diverse range of forces 'push and pull' the NUA from different angles. The chapter recommends enactment and adoption of stimulating laws and policies towards promoting sustainable cities.

Keywords Institution · Governance quality · Policy · Planning law · Proactivity · Urban economics

S. S. Mazongonda (✉) · A. Chigudu
Department of Architecture and Real Estate, University of Zimbabwe, Harare, Zimbabwe
e-mail: simbasho@gmail.com

C. Chavunduka and I. Chirisa (eds.), *New Urban Agenda in Zimbabwe*,
Advances in 21st Century Human Settlements,
https://doi.org/10.1007/978-981-97-3199-2_11

1 Introduction

Accessible sources concurred that urban planning laws are fundamental in creating and managing urban environs devoid of nuisance, noise and harm, yet urban planning laws of most African countries have failed to keep pace with the rate of urbanisation (Berrisford 2011; Africa Research Institute 2013). Specifically, Africa Research Institute (2013, p. iii) claimed that "...in most countries, outdated, inappropriate and unintegrated laws are exacerbating urban dysfunction". Upon the attainment of independence, most African countries inherited spatial planning laws enacted by colonial governments. The colonial planning laws were characterised by segregationist ideologies meant to propound power disparities between the social classes and domination of colonial forces (Chirisa and Dumba 2012). Some revisions and amendments to these laws have been made with the view of relaxing segregationist tendencies. Such revisions led to rapid migration of people from rural to urban communities. This rural–urban drift coupled with natural increase in population has led to sheer unmanageable urbanisation. It has been argued that unless urban areas can be made more sustainable, and rural life more tolerable, the legacy of negative environmental and social costs will become irreversible (Farmer et al. 2006).

New planning approaches characterised by inclusivity, resilience, sustainability, use of appropriate tools and responsiveness have been proposed as strategies for developing new towns and managing existing ones. One such approach is the New Urban Agenda (NUA). The NUA seeks to ensure a better and inclusive future by balancing out economic, social and environmental sustainability. The agenda acknowledges the complexity of urban environments as myriad land uses such as housing, industrial, recreational, institutional and commercial. This complexity was partly explained by Chaeruka (2002) who argued that planning law is theoretically and practically complex because it connects numerous competing dots in urban environs. The broadness, diversity and cross-sector visibility of planning law is something that can be likened to a blind man and elephant problem (Schneider 2005); everybody touches the elephant, yet only the part they would have touched. To have a full understanding of this subject and consequently inform policy well, this chapter provides a prescriptive analysis of laws and regulations that must be enacted and upheld in Zimbabwe if the NUA is to realise its stated and implied purpose. Arguably, success of the NUA is partly enabled by enactment and revisions of relevant spatial planning laws and regulations. This chapter first discusses the methodology used and then traces urban planning laws of Zimbabwe with view of demonstrating how and why they have been labelled outdated. This is tailed by a discussion on the NUA through the lens of urban planning laws. Then, a prescriptive discussion of how urban planning laws must be aligned to the NUA is given before drawing up conclusions.

2 Methodology

This chapter used both secondary and primary data with a greater inclination towards secondary data. Secondary data was gathered using literature review and documentary examination of administrative documents. Textbooks, journals, unpublished studies and conference papers were the main sources of literature. These sources helped in establishing what has been and what is yet to be studied and they also provided rich, historical and current scholarly information on planning law. Documents reviewed include statutory plans, parliament acts and statutory instruments (SI) that govern urban development in Zimbabwe because Zimbabwean planning law is explained and expanded in statutory plans, acts of parliament and SI. It was, thus, sensible to review these documents to assess their provisions. In addition, the NUA endorsed by the United Nations General Assembly (UNGA) on 23 December 2016 was reviewed with the view of understanding the proposed approach to urban development and management. This was essential since it enabled analysis of existing institutional framework in creating inclusive space for development.

Then, primary data was gathered from key respondents such as representatives from administrative court, high court, practising planners and private consultants. After reviewing accessible literature, planning laws, regulations, and SIs, planning practitioners were purposefully selected to give their expert opinion on current state of planning laws in Zimbabwe and prospects for aligning them with the NUA. Among the five selected practitioners were one High Court of Zimbabwe (HCZ) judge (who handled planning law cases between 2004 and 2010 when she was a judge at the Administrative Court of Zimbabwe (ACZ)), two selected town planning assessor with the ACZ, a private consultant specialising in planning law and urban development and a former Deputy Director of the Department of Physical Planning (DPP) in Zimbabwe. These planning experts were selected since they are exposed to planning related issues daily. As such, they have experience and expertise in handling and interpreting planning law. An interview guide designed using probing, follow-up and specifying questions was used in carrying out semi-structured interviews with planning law experts. This guide was developed using researchers' interpretation of existing planning laws, the NUA and related literature. Interview data was analysed using conversation analysis, and secondary data was analysed using thematic and content analyses.

2.1 Outdated Spatial Planning Law in Zimbabwe: A Review

Chaeruka (2002) historicised the development of planning law in Zimbabwe with much emphasis on development control. Chaeruka's study revealed that planning law in Zimbabwe has undergone five stages between 1900 and 1996 classified as follows: (1) case for development control (1900–1933); (2) inadequate planning law (1934–1945); (3) broadened base for planning law (1946–1975); (4) perfected planning law (1976–1990) and (5) challenged planning law (1991–1996). Informed by Chaeruka's classification, this section traces the development of Zimbabwe's planning law system from pre-independence to post-independence. It must be noted that Zimbabwe attained independence in 1980. As such, the pre-independence era covers the period before 1980 and the post-independence era covers the period from 1980 to date. This is aimed at establishing the origins, characteristics, and extent to which current laws have been outpaced by changing socio-economic environment.

2.1.1 Before 1980

The backbone of planning law is policy, acts of parliament, regulations and institutions governing the development of the built environment. During the early years of colonisation, development was at its minimum and there was no need for strict control of urban developments. With increased urbanisation and motorisation from mid-1920s, there arose the case for development control leading to the enactment of the first Town Planning Act (TPA) in 1933 (Government Town Planning Office (GTPO) 1951). The legislation was mainly applicable to land with title (that is, European areas and commercial areas) following the provision of the Land Apportionment Act of 1930 that divided land into urban land, commercial farms and communal land. According to the GTPO (1957) the first TPA dealt with issues inclined to zoning, disposal of solid and liquid waste, subdivision of land, protection of amenities and buildings and enforcement of planning control. The 1933 Act was mainly applicable to six urban areas of Harare, Bulawayo, Mutare, Gweru, Masvingo and Kadoma. Being informed by the Land Apportionment Act, the TPA had a colonial bias of planning serving white settler interests in European areas only.

Between 1934 and 1945 the GTPO was overwhelmed with spatial planning cases because of urban sprawl and migration of Europeans into Zimbabwe (then Rhodesia), especially after the Second World War (SWW). Settlements such as Waterfalls in Harare were established with inadequate services such as piped water, sewer, public transport facilities, schools, and shopping facilities (GTPO 1951, 1957). The two planning officers working from Bulawayo and Harare were overwhelmed with town planning cases and could not cope with the rate at which urban development was taking place. Furthermore, the TPA of 1933 did not provide detailed procedures and guidelines for the development of town planning schemes (town plans) and subdivisions of existing plots (Chaeruka 2002). It has been argued that the crisis of dealing with inadequate planning law led to the enactment of the Town and Country Planning Act (TCPA) of 1945 (Local Government Reorganisation 1974). The 1945 TCPA emphasised on clarifying processes and procedures for subdivision and development of town planning schemes. By then, the need for compact settlements was popularised for ease of serviceability and circulation. Just like the TPA of 1933, the 1945 TCPA only proposed revisions to be applied in European communities only. This is a true testimony that planning laws were only designed to serve the white settler community.

Having realised that planning instruments and institutions were outpaced by development, the colonial government set up the Capewell Commission of Inquiry (CCI) in 1949. It was mandated to review the spatial planning system with much emphasis on the planning framework in place at that time, challenges faced in operationalising it and opportunities it presented with view of mapping on the way forward. The CCI clarified and addressed numerous problems in the planning system with the view of balancing the private and public interests. The recommendations of the CCI saw major changes between 1946 and 1975. Among them were the:

1. Establishment of an endowment policy to enable Local Authorities (LA) to finance administrative activities. The endowments were to be charged on subdivisions.
2. Setting up of services provision policy to ensure that developers provide basic standardised public infrastructure such as roads and setting up of African settlements in European areas to provide labour in growing urban settlements.
3. Setting aside of servitudes in all layout plans. Servitudes were to be used by utility companies in laying out power, water, telephone and sewer lines.
4. Establishment of standard plot size in subdividing properties and the developers' liability in having such subdivisions, that is, the extent to which they must service the properties when subdividing.
5. Coming up with a green belt policy in 1955 meant to contain urban sprawl and protect agricultural land surrounding urban areas to ensure that urban areas can generate food to feed their populace.
6. Clarification of issues to be covered in outline and scheme plans because most provisions in these tools were outpaced by development, rendering them ineffective. The revisions of these plans were partly enabled by issues discussed in town planning cases (case law).
7. Revision of the subdivision policy considering the increasing number of subdivisions and complexities associated with them. Actors, such as representatives from Ministries of Health, Internal Affairs, Local Government, the Natural Resources Board, the Surveyor General and Deeds Registry and Departments of Water and State Roads were to be consulted before a subdivision is approved.
8. Consideration given to environmental issues, particularly air pollution by industries and land degradation due to sand extraction in urban peripheries. A reactionary allied act, Atmospheric Pollution Prevention Act (APPA) [Chapter 20:03], was passed in 1971 to control air pollution. Musandu-Nyamayaro (1993) noted that challenges experienced in the late 1960s and early 1970s ignited planning authorities to adopt mixed land use approach and expand vertically, thereby curbing urban sprawl and its associated challenges to the environment.
9. Observation that the use of scheme plans to control development led to problems of planning control outside scheme boundaries. It was also noted that the 1945 TCPA did not provide for planning control of African townships within European areas and communal lands.

The period 1946–1975 saw many planning issues, factors and actors brought to light in dealing with practical complexities of urban planning. These revisions in planning instruments are a true demonstration of the diversity and broadness of spatial planning because everything happens in space. Planning control policies, laws and regulations formed between 1946 and 1975 led to the promulgation of the Regional, Town and Country Planning Act (RTCPA) [Chapter 29:12] in 1976. Chapters of the 1945 TCPA were rearranged, flesh was added and new areas were also added to include the regional and rural component. One classical scholar argued that planning acts enacted up to 1947 were more of skeletal laws and needed flesh for them to be effective (deSmith 1948). Arguably, the RTCPA [Chapter 29:12] gave flesh to the skeletal acts. To a larger extent, recommendations made by the CCI saw light of the day. For example, the green belt policy was put into practice in 1955 and the APPA [Chapter 20:03] was passed in 1971 following the CCI recommendations. Furthermore, most of the recommendations were reflected in the RTCPA [Chapter 29:12] enacted in 1976.

Overall, the pre-independence era was characterised by segregationist tendencies and institutional overlap. Firstly, all policies and acts adopted before 1975 had a bias towards promoting order, health, safety, amenity, and convenience in urban areas only (formerly European towns). This observation was reinforced by Chirisa and Dumba (2012) who lamented that these laws were racially skewed with a bias towards promoting the welfare of the settler community only. This skewness was also spelt out by the presence of laws that did not provide for planning control of African townships within European areas. All needed was labour from Africans, without ensuring their health, safety, and order. Secondly, institutional overlap and legislative loopholes are some of the factors that led to inadequacy of planning law. For example, at least seven government departments were to be consulted before a subdivision is approved (see CCI recommendation in some pervious paragraph). This consultation process created loopholes for corruption, took a considerable amount of time for decisions to be made, and duplicated duties in certain instances.

2.1.2 Post 1980 to Date

Following the attainment of independence in 1980, the whole country was put under the planning control of the RTCPA [Chapter 29:12] enacted in 1976. This was meant to remove the discriminatory classification brought about by the Land Apportionment Act of 1930 (European-African areas separation). Up to 1990, a few town planning cases were brought before courts following the enactment of SI 379 and 380 of 1982. These SI clarified permitted developments. Another notable change was the rearrangement of local planning areas following the enactment of the Rural District Council Act (RDCA) [Chapter 29:13] in 1988.

Between 1991 and 1996, most of the changes were administrative and not procedural in nature. For example, merging of rural and district councils following the enactment of RDCA [Chapter 29:13]. This was meant to put Rural District Councils (RDC) at par with Urban Councils (UC) in terms of powers to prepare master, layout and local plans, and control development in their areas of jurisdiction. However, most RDC and small towns did not have the capacity to hire qualified planners. As such, the responsible minister, though the Department of Physical Planning (DPP), was made responsible to develop master and local plans on their behalf. Zimbabwe Institute of Regional and Urban Planners (ZIRUP) Papers (1990–1996) noted that this approach delayed the planning process. This deregulation process coincided with the adoption of Economic Structural Adjustment Programme (ESAP) in early 1990s. Hasty implementation of ESAP saw the chronic exodus of industrialists, a downward spiral of economic activities, privatisation of firms, among other notable changes (Zinyama et al. 1993). Landslide retrenchment of workers pressured them to find sanctuary in informal enterprise.

Further precipitating the progression of informal enterprise was the awarding of millions of dollars in compensation to ex-combatants in the late 1990s which left the country's economy in tatters (Mzumara et al. 2015). These changes led to relaxation of zoning laws partly explained by the gazetting of SI 216 of 1994 that provided for development of home industries in residential areas. Home industries have been defined by Kenton (2020) as small-scale, decentralised manufacturing businesses often operated out of a home, but within residential areas. Kuah (2002) added that home industries are a mere concentration of competing and complementing firms in a residential zone. The RTCPA [Chapter 29:12] that was promulgated in 1976 was revised in 1996, but the revisions did not lead to material changes. As such, to date, spatial planning issues in Zimbabwe are guided by the provisions of an Act that is at least 45 years old. Between then and now, a lot has changed in all spheres of urban life. This begs the question; is the current RTCPA fit enough to provide for and manage urban development we are experiencing today? Yet, classical scholars such as Kanyeihamba and McAuslan (1978) lamented that urban development takes place in a dynamic fashion, and planning laws that guide it must follow suit.

From the historical trace of the evolvement of planning law in Zimbabwe, it is amply clear that laws, regulations and policies are set up in a reactionary manner. Planning instruments are outpaced by development, and this has led to manifestation of numerous problems, some of which are being experienced in the present day. This is exacerbated by weak spatial planning institutions, especially in rural areas. As alluded to earlier in this chapter, planning law is a broad subject that can be viewed through economic, health, sociological, geographical, planning, business and environmental lenses. One of the fundamental objectives of this chapter is to prescribe revisions to existing planning laws with the view of partly enabling the realisation of the NUA. In one of the interviews with a HCZ judge, who is a former president of the ACZ, researchers were advised to directly quote statutes on point of law and then analyse them on point of fact. In her own words, she said:

> I strongly encourage you quote the exact provisions of statutes on point of law because paraphrasing has risk of misleading readers whenever you misinterpret the law. Even in courts, lawyers, prosecutors, and judges first quote relevant sections from statutes and then apply them to specific cases. This gives interested parties a real feel of law. In your case, tell us what the law says before applying it to what was observed and reported.

Following this advice by an expert in spatial planning law, all points of law in this chapter are based on direct quotations of relevant sections from statutes and then applied to specific cases. Since planning laws work hand-in-glove with allied Acts, relevant sections from the RTCPA [Chapter 29:12], Administrative Court Act (ACA) [Chapter 7:09], Roads Act (RA) [Chapter 13:18], Environmental Management Act (EMA) [Chapter 20:27], SI 216 of 1994 and the New Constitution of Zimbabwe are used in providing a real feel of the law before application. In adjudicating town planning matters, judges work with assessors as provided for by the Section 6 of the ACA [Chapter 7:09]. For this reason, one key informant (a HCZ judge, who presided over town planning cases between 2004 and 2010) referred the researchers to five assessors to help in assessing and interpreting observed and reported issues on point of law. Two of the assessors advised use of case of law in analysing issues with a greater inclination to planning law. In his own words, one of the assessors said:

> Every game has its own rules, formal and informal. In context of law, principles of certainty, justice and uniformity are some of the rules. To have a strong footing in debating an issue, use case law and legal language on point of fact. Case law gives you experiences elsewhere on similar, but not identical cases. As such, case law must be contextualised to specific cases on point of fact.

This advice was also seriously considered in argumentation. There was also general advisory consensus among all the town planning experts that the researchers obtain and utilise primary evidence including pictures and mapped data in undercutting arguments. As such, this chapter is based on direct quotes from statutes on point of law, contextual application of case law on point of fact and picture and mapping narrations in the analysis of the relationship between town planning and current realities in Zimbabwe. Where do current laws of planning in Zimbabwe take us regarding the realisation of the NUA? Perhaps, reviewing the NUA provides a point of departure in prescribing the refinement and alignment needed to existing laws.

2.2 *The NUA Perspective on Urban Laws*

One of the topical issues in development space is the NUA entrenched in Habitat III (2017). The agenda chats the envisioned approach to urban development in creating inclusive spaces and generating employment. The agenda is anchored on five pillars: (1) national urban policies; (2) urban legislation and regulations; (3) urban planning and design; (4) local economy and municipal finance and (5) local implementation (Habitat III 2017). Interestingly, planning law is instrumental in the realisation of the NUA since it provides for the creation of settlement structure, managing it and controlling non-provided for developments. Wekwete (1989) further argued that planning law provides the basis for taxation by setting categories for land use and derive land value therefrom. The definition and intent of planning law shows that it is the best vehicle driving the five pillars of the NUA.

It must be noted that urban policies, laws, and regulations inform the configuration of land uses in space and this speaks to the urban planning and design pillar of the NUA. Compact and mixed designs advocated for by the NUA reduce time and cost of travel from one activity to another, at the same time promoting efficiency and effectiveness in service delivery. Moreover, they help contain the negative effects of urban sprawl, something that the government of Zimbabwe has been fighting against since the colonial era. In pursuance of sustainable urban designs, the NUA calls for flexible, inclusive and just urban legislation and regulations. The next section discusses current realities in Zimbabwe regarding application of planning laws and provides a prescription for best practices if the NUA is to be realised as intended.

3 Discussion of Findings

3.1 Urban Informality

Having realised increasing informality from 1990 onwards, the Government of Zimbabwe (GoZ) enacted the SI 216 of 1994 to provide for inclusive manufacturing spaces within residential areas through Use Groups 4, 5 and 6. Three main manufacturers' clusters that organically grew in the capital city (Harare) were regularised. These are the Complex in Glen View, Gazaland in Highfield and Siyaso in Mbare (Mazongonda and Chirisa 2017). This noble reaction by the government has been overtaken by natural increase in population since no additional space has been provided for since the enactment of SI 216 of 1994. In the recent past, Mazongonda (2019) carried out a census survey of manufacturers working in the Complex, Gazaland and Siyaso and revealed how home industries have grown beyond their set boundaries. Figure 1 is a map showing the intensity and interaction of manufacturers working in Siyaso home industry.

The boundaries of Siyaso are shown by a polygon in blue. So, all pins outside that boundary represent outliers who are also into manufacturing but cannot be housed in Siyaso. Such outliers are manufacturers working along roadsides and servitudes, and some working along riverbanks of Mukuvisi River. Mukuvisi is one of the rivers that feed into the main source of water for Harare. The authorities carry out frequent crackdowns on these manufacturers, but they come back after such operations. Economic experts view these outliers as symbols of successful clusters, yet spatial planning experts view them as violators of planning provisions. Instead of cracking down on them, local planning authorities (LPAs) must secure additional land close to such clusters or elsewhere and construct manufacturing shells to house the outliers. More preferably, additional shells must be constructed within neighbourhoods of existing ones to promote compactness of settlements. The legal parameters governing the establishment and management of home industries must be revised to align with current realities. Overconcentration in space poses public health threats and further burdens the already overloaded reticulation systems. For example, in Fig. 1, some

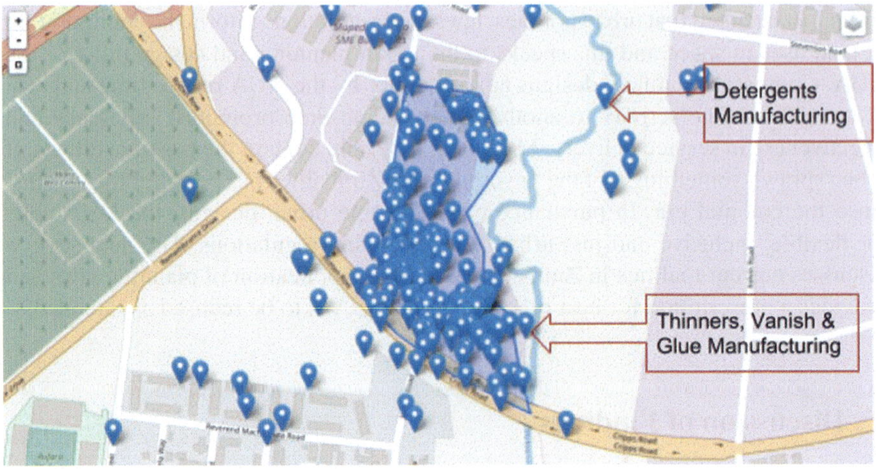

Fig. 1 Riverbank manufacturing in Siyaso (Mazongonda 2019)

manufacturers of detergents, glue and thinners have conglomerated along the banks of Mukuvisi River where they discharge some of their waste into the river.

3.2 Enumeration and Mapping Technologies

The preceding sub-section has shown how informal manufacturing activities in Harare have grown beyond their designated boundaries. Arguably, LPAs must not limit their action to removal, demolition and eviction as provided for by Section 35 of the RTCPA [Chapter 29:12]. Several sources have criticised frequent crackdowns on illegal developments by the GoZ (Tibaijuka 2005; Gumbo and Geyer 2011; Mazongonda and Chirisa 2017). It has been argued that such crackdowns give new life to old conceptions since they are informed by statutes that are not in consonant with current realities (Tibaijuka 2005). Scholars such as Varley (2013) have argued that informality must be treated as the identity of developing cities. LPAs are encouraged to view growing informality through the lens of the goal of humanising cities. For example, Section 13(1)(b) of the RTCPA [Chapter 29:12] provides that:

> …before altering or replacing any master plan or local plan, (a local planning authority shall) must undertake, if it thinks fit, a fresh study of the planning area or any part thereof…

Applying this provision to the gigantic force of urban informality, it is sensible to undertake fresh research of the planning area to determine the magnitude of informality (in this case) and consider redeveloping the area to accommodate the growing population. Such fresh studies must be supported by technologies of mapping and enumeration to come up with accurate and authentic results to inform policy. This is against the background that some spatial planning decisions are made without accurate and up-to-date data. For example, Roy (2009, pp. 80–81) explained the significance of mapping as follows:

> …while it has been often assumed that the modern state governs its subjects and conducts planning though technologies of visibility, counting, mapping, and enumerating, in previous work I argue that regimes of urban governance also operate through an 'unmapping' of cities.

Roy (2009) used evidence from India to explain how poor governance and weak institutions have exacerbated informality. Government must set up a policy that makes it compulsory to use technologies of enumeration and mapping when carrying out fresh studies to inform operationalisation of master and local plans. Such a policy must set the parameters for use of geo-spatial intelligence in carrying out fresh studies, designing development plans, capacity building for creation of strong institutions and managing existing settlements. Moreover, use of modern technologies such as use of geo-spatial intelligence to manage settlements makes everything transparent for all to see, thereby eliminating tendencies of tampering around with city and property boundaries.

3.3 Proactivity Versus Reactivity

Evidence from the historical trace of the evolvement of planning law has shown that it is reactionary in nature. Under ideal conditions, planning law must be on the lead in setting the parameters and practice of urban development. Chaeruka (2002) has argued that the reactionary nature of planning law is meant to close gaps noted in its practical application. This is so because planning law is broad, diverse and dynamic. The current RTCPA [Chapter 29:12] was last revised in 1996, but a lot has changed in the development space between then and now (Chigudu 2021). A planning commission comprised of experts drawn from urban economists, planning lawyers, town planning assessors, prominent lecturers in planning, representatives from LPAs and the DPP must be set to deliberate on the future of spatial planning in Zimbabwe. In addition, experts in demography and settlement, actuarial science and geo-spatial intelligence must have seats in such a commission. They will play an instrumental role in estimating, using mathematical models, future population trends, demand for housing and transport infrastructure and related issues. They must sit at least once a year and chat on the direction of spatial planning for the next two or three years since planning issues are dynamic. Traditionally, commissions were set to react to challenges noted. For example, the CCI was set in 1949 to review the inadequacy of planning system in place at that time. On the contrary, commissions of planning must be proactive on planning issues. Instead of spending resources on evicting, demolishing and removing illegal developments, such resources must be used in planning what the future should be like considering current trends.

One practical case drawn from transportation and traffic planning is a clear testimony that some of the challenges we are experiencing today are emanating from the fact that LPAs are not investing in proactive planning. The Traffic and Transportation Study of 1984 conducted by the Swedish International Development Cooperation Agency (SIDA), the Harare Combination Master Plan (HCMP) of 1993 and the Harare Traffic Management Local Subject Plan of 1997 emphasise the importance of using research to inform policy and practice. Proposal Number 5 of the Local Subject Plan reads, '*a detailed transportation study should be undertaken using an appropriate model*'. Obviously, an appropriate model is one that is amenable to traffic patterns in Zimbabwe, and that traffic pattern is continually changing due to increase in motorisation. Further, the HCMP calls for review and recommendations after a 20-year period, to factor in changes. A mathematical deduction indicates that 2013 was the year for review. Dumba (2019) lamented that up-to-date calls for such studies are yet to be taken up. This partly explains why Harare, the capital city of Zimbabwe, is experiencing congestion challenges because they are using outdated models.

3.4 Urban Economics and Urban Planning Interface

Historical evidence has shown that the CCI of 1949 was composed of numerous professionals excluding urban economists. However, it has been argued that urban economists must have a seat in any planning commission because they better understand the dynamics in the land market (Bertaud 2018). In a largely free market economy such as Zimbabwe, forces of demand and supply must be allowed to determine prices of land and subsequently determine the pattern of land use. For example, the invasion of residential areas surrounding the Central Business District (CBD) of Harare by commercial uses is a true testimony that market forces can determine the pattern of land uses. Imagine a commercial facility that intends to occupy an area set aside as a residential zone and is willing to pay a higher price and compensate affected residents so that they relocate to another place. This will be mutually beneficial because the large commercial facility will significantly contribute to fiscus, residents will have extra money after being paid a price higher than the current value of their properties and the commercial investors will realise more profits because of their convenient location. However, Mazongonda and Muromo (2011) revealed that some commercial activities in Belgravia, Harare are regarded as illegal and are subject to planning control, yet they are contributing meaningfully to fiscus. Bertaud (2018) lamented that planners are informed by rigid laws of planning, in this case the RTCPA [Chapter 29:12], which is regarded as a planning bible with holy authority to sanction what is legal and illegal, and what is permitted and not permitted. Processes of regularisation, determination of planning permission applications and appeals must be informed by urban economic principles since they are informed by and respond to market forces. Currently, Harare's CBD is growing beyond its set boundary, encroaching into the central district, in terms of development control, influencing change of use in areas such as Eastlea, Belgravia, Milton Park, Avondale and Highlands. As such, responsive planning, informed by economic principles, must revise and relax provisions for giving special consent to land uses not fit to be erected in certain zones. This is so, because the planning process has been labelled as long and winding, rigid and time-consuming (Chaeruka 2002).

3.5 Rooting Up Aspects of Delay

A study by Chaeruka (2002) brought to light inefficiencies in the Zimbabwean planning system using empirical data from ACZ files. Chaeruka's study used the time taken to process planning permission applications at the LPA, time spent between LPA decision and planning permission application and lodging of notices of appeal, and time taken to determine planning appeals at the ACZ using 74 sampled cases deliberated on between 1976 and 1996. An average of 90 days (3 months) for each of the three aspects of delay were established. Overall, approximately 9 months of productive time is lost in deliberating on town planning cases. From an economic viewpoint, time is money and a minute lost is never gained. Regardless of the outcome of the decision, all parties involved would have lost revenue. The argument here is not to give town planning applications and appeals a predetermined outcome, but to reduce the number of days in processing such applications. As argued earlier, applications must be processed through an economic lens. Just imagine an application being approved after at least 90 days, yet forces of demand and supply are ever dynamic. Arguably, at the time of approval, a lot would have changed, and the applicant will be looking for a new economic opportunity that requires another at least 90 days to be approved. Such a planning environment discourages innovation and investment.

3.6 Rooting Up Elements of Corruption

In general, the term *'corruption'* refers to the misuse of resources or power for private gain. Kolstad et al. (2008) defined it as the abuse of entrusted power for secluded gain. This chapter is premised on the distinction between political and bureaucratic corruption. Political corruption takes place at the highest levels of political authority (Andvig et al. 2001). It involves politicians, government ministers, senior civil servants and other elected, or appointed senior public office holders. Interviewed key informants and urban citizens confirmed reports and public outcry of corruption cases by public officials in Zimbabwe. Political corruption is the abuse of office by those who decide on laws and regulations and the basic allocation of resources in a society (that is, those who make the *'rules of the game'*). The courts and other spatial planning institutions confirmed cases of tailoring laws and regulations to the advantage of private sector agents in exchange for bribes, granting large public contracts to specific firms or embezzling funds from the treasury. Both political and bureaucratic corruption are common in planning institutions in Zimbabwe.

Interviewed planning experts further confirmed that poor working conditions have magnified cases of bribes, embezzlement, collusion, fraud, extortion and patronage in both rural and urban local authorities.

Grand corruption that includes huge scale of corruption and the considerable sums of money involved is rampant at LPA level. Evidence gathered has revealed that bureaucratic corruption occurs during the implementation of public policies. This involves appointed bureaucrats and public administration staff at both central government and local authority level (Kolstad et al. 2008). There have been cases of corrupt acts among those who implement the rules designed or introduced by top officials, including transactions between bureaucrats and with private agents (for instance, contracted service providers). Such agents may demand extra payment for the provision of government services; make speed money payments to expedite bureaucratic procedures; or pay bribes to allow actions that violate rules and regulations.

Interviewed town planning experts also confirmed cases of payment or taking of bribes or kickbacks to obtain posts or secure promotion, or the mutual exchange of favours in most public institutions including local authorities. There is evidence that corruption at the top of a bureaucratic system increases corruption at lower levels (Chand Sheetal and Moene 1999; Kolstad et al. 2008). Corruption is closely linked to the generation of economic rents and rent-seeking. This refers to actors securing above-normal returns from an asset; not by adding value to it through investment but rather through manipulating the social and political environment. Rent-seeking involves corruption whereby the payment of bribes is necessary to manipulate the environment to benefit a particular actor. Evidence reviewed has shown that in Zimbabwe, corruption is exacerbated by weak governance systems and institutions. There are also particular challenges related to the measurement of corruption. This owes to the clandestine nature of corruption and the reliance of corruption measures on perception-based data, which themselves are determined by understandings of corruption that vary across countries and societies (Kolstad et al. 2008).

Smoothness of planning system is a function of reducing or totally eradicating corrupt tendencies among everyone involved in the planning process. Corruption is more of a cancerous tissue that is difficult to cure. Using the case of Indian cities, Roy (2009) argued that politicians and regimes of urban governance also operate through corrupt channels to extend their tenure in office. This is type of corruption is not peculiar to India. Murowe and Chirisa (2006) explained how politicians tolerated formation of housing cooperatives in Harare and allowed them to occupy semi-serviced pieces of land. Most of these cooperatives have been named after heroes and heroines that liberated Zimbabwe from colonial rule. For example, Joshua Nkomo Heights, Sally Mugabe Heights, and Josiah Tongogara housing scheme. Residents of these newly established residential areas are not given title to the land they occupy so that they remain loyal to the ruling party, Zimbabwe African Nation Union-Patriotic Front (ZANU-PF). Residents' tenure of occupation is dependent on their loyalty to the so-called *'bigman'* (political heavy weights).

3.7 *Preservation of the Environment*

Urban landscape structures that increase liveability for people also play an important role in sustaining wildlife population. Towns and cities are often built in the vicinity of features that are important to natural vegetation and wildlife habitats that include confluence of rivers and forests. Green spaces in urban areas significantly contribute to sustainable development (Atiqul Haq 2011). While urban development can disturb larger habitat areas, urban areas can contain key natural areas and features that offer significant benefits to the natural environment and wildlife. The role for urban ecosystems in wildlife conservation has become increasingly recognised in recent decades (Atiqul Haq 2011; Panagopoulos et al. 2015). Through deliberate spatial planning approaches, public green spaces are set aside in urban areas to engage people with nature and enable citizens to enjoy the outdoors where they live and work. Issues of habitat conservation and restoration should be given priority by urban residents.

The urban areas are branded by the frequency of built-up structures and impervious surfaces that discourage surface water flow and disturb water quality, reduce vegetation cover and diversity and cause habitat loss, destruction and degradation (Panagopoulos et al. 2015). Human activities are concentrated in cities and towns. These activities displace wildlife, produce pollutants, noise, heat and artificial lighting that can disturb the natural environment. Cities and municipalities are increasingly working to decrease many of these problems through enforcement of environmental protection laws and adoption of best practices (Atiqul Haq 2011; Panagopoulos et al. 2015). However, in some instances structures are built on wetlands disregarding the requirements for Environmental Impact Assessments (EIAs). This is promoted by political interference and patronage practices.

4 Conclusion and Recommendations

This chapter has provided a prescriptive analysis on how laws, policies and institutions must be structured towards the realisation of the NUA. It has been argued that most laws and policies of planning are outdated and are no longer in consonant with current realities. Examples, drawn from urban informality, urban economics, enforcement of planning control, governance quality, environmental conservation, technological factors and planning processes have been used to suggest the parameters for achieving the goals of the NUA. Considering the broadness and diversity of development, stimulating regulations must be put in place. Against this background, Schneider (2005) has likened theoretical complexity and practical controversy of planning law to *a 'blind man and elephant problem'*. It is broad to the extent that it must be approached from different disciplinary perspectives. This is so because planning law can create either an enabling or inhibiting environment in all spheres of urban life. Under ideal conditions, planning law must reflect and assert the public interest without violating private interests. Prescriptions given in this chapter are indicative and not conclusive in nature, but their totality seeks to balance between public and private interests. Therefore, it is recommended that:

- Home industries must be expanded to accommodate manufacturers working outside set boundaries. Policies designed by government must consider the traits of organisation among the manufacturers which can be nurtured, improved and carved to ensure sector accountability and spatial orderliness. The idea of inclusivity can be applied in the wake of growing urban informality. This can be worked out through inclusive design and adaptive planning which is receptive to the practical requirements of the sector.
- A planning commission must be set to sit at least once annually to discuss the future of planning rather than setting commissions that react to challenges already experienced. The GoZ, through the DPP, must set aside financial resources to smoothen the hosting of such an annual event. Forward-thinking policy formulators and other participants need to chart a new course that enables envisioning of the future and how existing laws, and policies can be carved to sync with the direction of spatial, economic, social and political change.
- Among other experts that will be part of the planning commission, urban economists must have a seat to share their experiences and thoughts regarding land values and land markets. Furthermore, experts in demography, actuarial science and statistics must be part of such a commission to help with estimation of future population trends, rate of motorisation and urbanisation among other parameters.
- Mapping and enumeration technologies must be used in carrying out studies in the urban and rural environs to provide accurate and reliable results for use in shaping policy and practice. The RCTPA, through its Section 13(1)(b), provides for carrying out fresh studies to inform redesigning of master plans but do not suggest how best that can be done. Such sections must be revised to emphasise the need for use of geo-spatial intelligence for improved quality of results.
- Times for processing planning permissions' applications and processing of appeals at the ACZ must be reduced with the view of improving efficiency and reducing cost of delay.
- All stakeholders in the development space must contribute towards reducing corrupt tendencies since corruption is being experienced at all levels of the development process. This is best achieved when the root causes of corruption are eliminated in planning processes. For example, regularly changing people sitting in positions of authority in government departments and LPA and making the process of awarding tenders more transparent through holding public debates to allow prospective applicants justify their proposals. Such debates must be adjudicated by members of a planning commission that are shuffled from time to time.

References

Africa Research Institute (2013) How to make planning law work for Africa. Counterpoint Series. Africa Research Institute

Andvig JC, Fjeldstad OH, with Amundsen I, Sissener T, Soreide T (2001) Research on corruption: a policy-oriented survey. Report No. R2001:7. CMI, Bergen

Atiqul Haq SMd. (2011) Urban green spaces and an integrative approach to sustainable environment. J Environ Prot 2:601–608

Berrisford S (2011) Revising spatial planning legislation in Zambia: a case study. Urban Forum 22(3):229–245

Bertaud A (2018) Order without design: how markets shape cities. MIT Press, Cambridge

Capewell Commission of Inquiry (CCI) (1949) Reports of the commission of inquiry on town planning and country planning control. GoZ. Southern Rhodesia

Chaeruka J (2002) Parameter and practice of development control in Zimbabwe: an evaluation of the system of development control with emphasis on the determination of planning permission and appeals. Unpublished MPhil Thesis. Faculty of Social Studies: University of Zimbabwe

Chand Sheetal K, Moene KO (1999) Controlling fiscal corruption. World Dev 27(7):1129–1140

Chigudu A (2021) The changing institutional and legislative planning framework of Zambia and Zimbabwe: nuances for urban development. Land Use Policy 100(1):1–10

Chirisa I, Dumba S (2012) Spatial planning, legislation and the historical and contemporary challenges in Zimbabwe: a conjectural approach. J Afr Stud Dev 4(1):1–13

deSmith SA (1948) Town and Country Planning Act, 1947. Mod Law Rev 11(1):72–81

Dumba S (2019) Modelling signalised intersections' capacity under the impact of minibus public transport in Harare, Zimbabwe. Unpublished DPhil Thesis. Faculty of Social Studies: University of Zimbabwe

Farmer P, Frojmovic M, Hague C, Harridge C, Narang S, Shishido R, Siegel D, Taylor P and Vogelij J (2006) Reinventing planning: a new governance paradigm for managing human settlements. A position paper developing themes from the draft Vancouver declaration for debate leading into the world planners congress, Vancouver 17–20 June 2006

Government of Zimbabwe (GoZ) (1991) A framework for economic reform: 1991–1995. GoZ, Harare

Government of Zimbabwe (1994) Statutory Instrument 216 of 1994 Regional, Town and Country Planning (Use Groups) Regulations, 1994 (Act 22 of 76). Government of Zimbabwe, Harare

Government of Zimbabwe (1996) Regional, Town and Country Planning Act [Chapter 29:12]. Government of Zimbabwe, Harare

Government of Zimbabwe (2001) Roads Act [Chapter 13:18]. Government of Zimbabwe, Harare

Government of Zimbabwe (2002) Environmental Management Act [Chapter 20:27]. Government of Zimbabwe, Harare

Government of Zimbabwe (2004) Design manual. Department of Physical Planning (DPP). Government of Zimbabwe, Harare

Government of Zimbabwe (2011) Administrative Court Act [Chapter 7:09]. Government of Zimbabwe, Harare

Government of Zimbabwe (2014) Zimbabwe's Constitution of 2013. Government of Zimbabwe, Harare

Government Town Planning Office (GTPO) (1951) Bulawayo peri-urban areas development plan. GoZ

GTPO (1957) Selected papers. Government of Zimbabwe

Gumbo T, Geyer M (2011) Picking up the pieces: reconstructing the informal economic sector in Bulawayo, Zimbabwe. J Town Reg Plan 59:53–64

Kanyeihamba GW, McAuslan P (eds) (1978) Urban legal problems in Eastern Africa. Uppsala: Scandinavian Institute of African Studies [Nordiska Afrikainst]. International Center for Law in Development, New York

Kenton W (2020) Article on cottage industry. Available online: https://www.investopedia.com/terms/c/cottage-industry.asp. Accessed 7 July 2021

Kolstad I, Fritz V, O'Neil T (2008) Corruption, anti-corruption efforts and aid: do donors have the right approach? Working Paper No. 3. Good Governance, Aid Modalities and Poverty Reduction: Linkages to the Millennium Development Goals and Implications for Irish Aid. ODI and CMI, London and Bergen

Kuah ATH (2002) Cluster theory and practice: advantages for the small business locating in a vibrant cluster. J Res Mark Entrep 4(3):206–228

Local Government Reorganisation (LGR) (1974) Report on LGR. Government of Rhodesia. Salisbury

Mazongonda SS, Chirisa I (2017) Order in informal enterprise: lessons from Gazaland Home Industry, Harare, Zimbabwe. Public Admin Dev Altern 2(1.1):16–31

Mazongonda SS, Muromo F (2011) Enforcement orders in Harare: tools for promoting development control. J Humanit (Zambezia) 33(i/ii):51–56

Mazongonda SS (2019) Urban informal manufacturing through the lens of planning law in Zimbabwe: case of Harare. Unpublished MPhil Thesis, Faculty of Social Studies: University of Zimbabwe

Murowe SR, Chirisa I (2006) Sally Mugabe heights: the tale of an incidental Zimbabwean peri-urban housing scheme seeking legitimization. Zambezia XXXIII(i/ii):79–89

Musandu-Nyamayaro O (1993) Housing design standards for urban low-income people in Zimbabwe. Third World Plan Rev 15(4):329–354

Mzumara M, Matthee M, Steenkamp E (2015) How a focused approach to export promotion can accelerate Zimbabwe's post-conflict reconstruction. Africagrowth Agenda 12(1):4–7

Panagopoulos T, Gonzalez JA, Dan MB (2015) Urban planning with respect to environmental quality and human well-being. Environ Pollut, 1–8

Roy A (2009) Why India cannot plan its cities: informality, insurgence, and the idiom of urbanization. Plan Theory 8(1):76–87

Schneider F (2005) Shadow economies around the world: what do we really know? Eur J Polit Econ 21(3):598–642

Tibaijuka AK (2005) Report of the fact-finding mission to Zimbabwe to assess the scope of operation Murambatsvina, UN special Envoy on human settlements issues in Zimbabwe. United Nations, New York

UK Department for International Development (2015) Why corruption matters: understanding causes, effects and how to address them. Department for International Development, London

United Nations Conference on Housing and Sustainable Urban Development (Habitat III) (2017) New urban agenda. Hibatat III Secretariat. Quito

Varley A (2013) Postcolonialising informality? Environ Plan d: Soc Space 31(1):4–22

Wekwete KH (1989) Planning laws for urban and regional planning in Zimbabwe: a review. RUP Occasional Paper No. 20. University of Zimbabwe, Harare

Zimbabwe Institute of Regional and Urban Planners (ZIRUP) (1990–1996) ZIRUP annual school conference reports

Zinyama L, Tevera D, Cumming S (eds) (1993) Harare: the growth and problems of the city. University of Zimbabwe, Harare

Climate Resilience and the New Urban Agenda in Zimbabwe: The Role of the Built Environment Disciplines and Practice

Charles Chavunduka, Patience Mazanhi, and Molly Kembo

Abstract In Zimbabwe, rapid urbanisation is sprouting informal settlements which are vulnerable to climate change. Climate change has been affecting the built environment through extreme weather events and disaster occurrence. It is important to address the impact of extreme weather events through resilient built environment fostered by both practice and curricula. To achieve this, the study relied on literature review, central and local government resilience policies, strategies and remote sensing data on disaster vulnerability. The findings show high levels of vulnerability in informal settlements which have been subject to flooding, environmental degradation and climate-induced diseases. They further reveal how Zimbabwe can use the New Urban Agenda to advance existing national strategies around climate resilience and the built environment. The study offers a framework for a resilient built environment in Zimbabwe.

Keywords Climate · Built environment · Vulnerability · Adaptation · Disasters · Resilience

1 Introduction

Africa is experiencing a rapid growth rate of its cities and this carries with it multiple challenges of housing shortages, poor sanitation and high risk to climate change hazards which worsen their vulnerabilities to disasters (Thondhlana et al. 2021). Challenges as these are not adequately solved by a single discipline, but demand the active contribution of both practice and the curricula. This means embracing of

C. Chavunduka (✉)
Faculty of Engineering and the Built Environment, University of Zimbabwe, Mt. Pleasant, Box MP 167, Harare, Zimbabwe
e-mail: cmchavunduka@yahoo.com

P. Mazanhi · M. Kembo
Department of Architecture and Real Estate, University of Zimbabwe, Mt. Pleasant, Box MP 167, Harare, Zimbabwe

C. Chavunduka and I. Chirisa (eds.), *New Urban Agenda in Zimbabwe*,
Advances in 21st Century Human Settlements,
https://doi.org/10.1007/978-981-97-3199-2_12

transdisciplinary mapping measures of climate resilience and sustainability in the growing cities and one of which is to integrate academic disciplines together with the existing societal climate knowledge (Opitz-Stapleton et al. 2010). Built environment disciplines play a significant role in promoting sustainability and the New Urban Agenda and this calls for inter-disciplinary and multi-disciplinary approaches to climate disasters (Lewis et al. 2005). This aids in creating a sustainable base for the built environment and human safety as disciplines bring together expertise which can be cascaded to communities towards urban climate resilience (Thondhlana et al. 2021). There is, but a limited amount of literature in regard with the relevance of built environment disciplines in pushing forward for the attainment of the New Urban Agenda and climate resilience. The chapter thus intends to bridge this gap through emphasising the importance of a climate resilience framework that provides a foundational base towards instilling climate resilience through disciplinary action.

The impacts of climate change on human settlements, particularly the vulnerable informal settlements are unavoidable. Tyler and Moench (2012) highlight that many built environments are exposed to dangers of climate change risks and this makes climate change adaptation and resilient strategies of importance. In Zimbabwe, rapid urbanisation is sprouting informal settlements which are vulnerable to climate change. Climate change has been affecting the built environment through extreme weather events and disaster occurrence. It is important to address the impact of extreme weather events through resilient built environment fostered by both practice and curricula. The study is long overdue to the incessant increase in the number of vulnerable people and the worsening climate events. According to Valencia et al. (2021), most of development resilient strategies are a spatial leaving a dearth of information on the synergy between climate resilience and the built environment. The chapter thus considers climate resilience in the spatial context of the built environment with the application of the climate resilience framework whose lens provide various pathways through which resilience can be integrated in urban development.

To achieve this, the study relied on literature review, central and local government resilience policies, strategies and remote sensing data on disaster vulnerability. The findings show high levels of vulnerability in informal settlements which have been subject to flooding, environmental degradation and climate-induced diseases. They further reveal how Zimbabwe can use the New Urban Agenda to advance existing national strategies around climate resilience and the built environment. The chapter offers a framework for a resilient built environment in Zimbabwe which shows the relevance of a multi-disciplinary and inter-disciplinary approaches to climate resilience in the built environment system. It answers the question on 'how have the built environment disciplines and practice been contributing to resilient built environment in Zimbabwe?' The study's organisation includes the theoretical framework which provides theoretical perspectives on urban climate resilience, the literature review that explores various aspects and experiences of other nations in regard with climate resilience in order to appreciate the relevance of climate resilience from a global to a local sphere. The chapter further on provides the methodology that justifies the study research and the results from the Zimbabwe's experience and finally the discussion and the conclusions.

2 Theoretical Perspectives/Framework

The built environment is a complex system that is classed in various scales from neighbourhoods, cities, regions and beyond. It includes manmade structures such as buildings and other infrastructure aspects which anchor around physical, economic, cultural and even social capital (Hassler and Kohler 2014). With the continuity nature of the built environment, urban climate resilience provides a comprehensive way to meet evolution of the built environment in the long run. This is initiated through exploring the changing conditions, which to a significant extent are being driven by climate change and this means providing relevant approaches through design, planning and governance. Resilience as a modelling and conceptual framework therefore carries within it phenomena that inhibit or enhance the attainment of sustainability goals and the New Urban Agenda. It has three main definitive dimensions as highlighted in the framework in Fig. 1. These include firstly the core definition of the resilience term, secondly the models that are generalised. This in simpler terms is a translation of the core definition into specific situations. Thirdly, the metaphors that show values and perceptions in regard with climate science.

A resilient built environment system is one that has the capability to foster continuity in spite of climate stresses and shocks (Al-Humuisqani and AlGhamdi 2022), while at the same time transforming and adapting towards sustainability and the New Urban Agenda. Of the most vulnerable built environments are the informal settlements whose location and even built environment renders them more vulnerable to climate disasters. A focus on climate change in such settlements demands a nuanced,

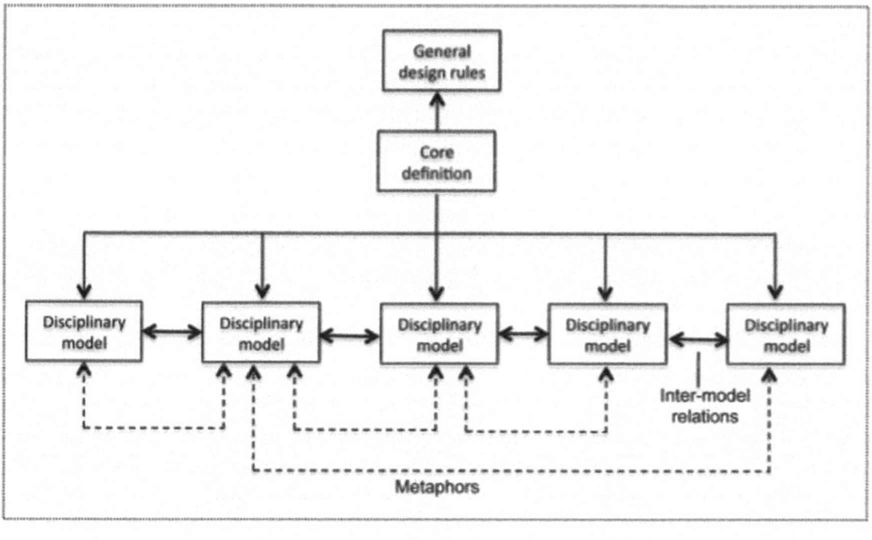

Fig. 1 Built environment disciplines-climate resilience framework. *Source* Hassler and Kohler (2014)

in-depth understanding of the vulnerability context of the locality. This is best shaped by the climate resilience framework and enables easier planning of interventions with the collaboration and integration of all built environment disciplines. Resilience is an aspect of sustainability (Home 2021), and requires a good governance approach that promotes community engagement in implementation as it addresses more than just the physical aspects but a whole socio-political and economic system. The vulnerability context of the informal settlements to climate change would demand the upgrading of housing, infrastructure and enhancing the capacity of the communities. It is imperative to have citizens and community members play a key role in participating for climate resilience. In order to achieve this, built environment disciplines play a key role in imparting climate knowledge that can be easily transferred to communities through practice.

Climate change, though being faced globally, is placing developing countries in a more vulnerable state despite them being faced with multiple challenges already. In order to foster resilient strategies locally, the application of a resilience framework is crucial in enhancing the practicability of resilience strategies on the ground. The climate resilience framework is described by USAID (2014) as a comprehensive tool that brings together elements, which when combined, provide the desired development outcomes followed from planning and implementation of resilient strategies. According to Beringer et al. (2019), the framework builds on resilience as the focal development aim of the built environment amidst climate change impacts. Figure 1 provides a climate resilience framework in relation with the built environment disciplines.

As highlighted in the framework, the core definition of climate resilience implies the general definition and understanding of climate resilience as a broad theme. That is, the ability of the built environment to bounce back even better after experiencing climate shocks and stressors such as disasters. This is followed by the model definitions that narrow down the understanding of climate resilience to fit into each respective discipline and its values. The existing metaphors as highlighted by Brand and Jax (2007) indicate how two or more disciplinary definitive models are connected in a transdisciplinary manner. The framework also shows the existing disciplinary relations which support the idea that the built environment is a complex system that needs comprehensive inter-disciplinary and multi-disciplinary approaches in tackling climate change especially in view of urban sustainability and resilience as supported by the New Urban Agenda.

However, it is important to assess the extent to which built environment disciplines are inter-disciplinary when it comes to climate change resilience. According to Haigh and Amaratunga (2010), the built environment has been considered from a traditional definitive perspective where the professions were more inclined to property and construction. The increasing challenges of climate change have called for the need to imbed climate into the disciplines if sustainability, through the New Urban Agenda, is to be attained. Apart from the highlighted inter-disciplinary nature, Chynoweth (2006) emphasises multi-disciplinary approach in which all the built environment disciplines are approached simultaneously in the context of the climate disaster circumstances to ascertain their relationship. No discipline is superior to another and neither does climate resilience come to fruition with only one discipline in action. Therefore, a disciplinary integration is necessary through the condition of pluri-disciplinarity as suggested by Haigh and Amaratunga (2010). This means a juxtaposition, made deliberately on the built environment disciplines to enforce their relationships. Coordination of the disciplines is the resultant effect thereafter promoted through cross-disciplinary action.

3 Literature Review

According to The United Nations Population Fund (UNFPA) (2018), at least 55% of the world's population reside in urban areas and are expected to increase to at least 60% by the year 2030 and up to 70% by the year 2050 (UNHABITAT 2014). Of the estimated populations, approximately 1 billion of the population reside in informal settlements, slums to be specific as highlighted by the USAID (2018). Of these, in sub-Saharan Africa alone, approximately 59% of the urban population live in slum settlements when compared with Latin America and Asia which have 21% and 28%, respectively. Already, these people suffer from lack of primary services of sanitation, housing and water and this is complicated by the emergence of climate change whose impacts are destructive (Satterthwaitte et al. 2020). This has left many communities so much devastated and trapped in a poverty cycle. Climate adaptation alone, is not enough without resilience to achieve sustainable development and the New Urban Agenda in the long run.

Home (2021) highlights that resilience has become a popular issue especially in regard with land management as emphasised by the Sendai Framework for Disaster Risk Reduction (2015–2030). Climate resilience in human settlements can be explained as the ability and capacity of the community to retain back its original state after withstanding climate stresses and shocks (Valencia et al. 2021). This means being able to adapt and adjust to the changing climatic environment. Often times, climate change shocks leave communities in very poor states than they were before. The climate-induced vulnerability worsens poverty as the settlement becomes susceptible to hazards (Al-Humuisqani and AlGhamdi 2022). But, when resilience has been fostered, it means that at the attack of climate change shocks, the communities are able to bounce back much better than they were before. Cities have a burden to ensure the implementation of disaster response strategies that build communities' capacity to face disasters (Tyler and Moench 2012).

Climate change disasters continue to affect human livelihoods and communities and thus call for urgent development action for risk reduction. Various conferences and protocols have been set up by various states in order to find effective ways to ensure climate resilience development. These include the Global Sustainable Development Goals' goal number 11 and 13. According to Valencia et al. (2021), goal 13 puts emphasis on the need to strengthen resilience through promoting awareness in development planning. Climate resilience cannot be separated from sustainable livelihoods and urban development as supported by goal 11 that puts focus on ensuring inclusive urban environments that protect the poor from extreme climate events (Home 2021). The New Urban Agenda considers the issue of climate resilience of huge importance as it highlights the need to ensure safe human settlements with low vulnerability to climate change risks while at the same time being highly responsive and resilient (UNHABITAT 2019). The New Urban Agenda makes the issue of resilient and sustainable urban development a prime factor as it is part of its three main transformative commitments to implementation.

3.1 International Treaties on Climate Change

Various international treaties and agreements have been made in order to fight against global climate change (USAID 2018). These include the Global Sustainable Development Goals (SDGs) which are governed by 17 goals that aim at improving human development through ending poverty and planet protection. In particular, goal number 11 and 13 put a major focus on promoting resilience and safe habitation for humanity by the year 2030 with no city left behind. The New Urban Agenda, the result of the United Nations Conference on Housing and Sustainable Urban Development that was held in Quito 2016 (UNHABITAT 2016), promotes city sustainability and inclusivity. According to Tollin and Hamhaber (2017), it is built on three major objectives and one of which emphasises on the need to create resilient and environmentally sound urban development.

Also, there is the Paris Agreement driven by the United Nations Framework Convention on Climate Change (UNFCCC) whose aim is to ensure the lowering of the average rise in temperatures to less than 2 °C as well as limiting their rise. The signatories to the Paris Agreement are guided by the Nationally Determined Contributions (NDCs) which are a must-have under the Paris Agreement. Zimbabwe as a member country, is listed among the lowest carbon emitters in the world, but one of the most vulnerable countries to climate change and thus, it is committed to lower greenhouse gases by at least 40% by the year 2030. Moderating temperature levels has a multiple effect of reducing climate change impacts. In addition, there is the Sendai Framework that focuses on lowering disaster risks and losses with a more urban-specific promotion driven by city-resilience motto (USAID 2018).

Brought together, the multiplicity of these international agreements prove to the fact that climate change is a real issue that needs global to local efforts in order to counter its impacts on human lives especially in urban areas (Home 2021). For the informal settlements, their locations, nature and even the built structures, place them in a more vulnerable condition of exposure to climate change hazards (USAID 2018). The locations are prone to landslides and flooding as they are found on river banks, hill slopes, garbage dumping sites and floodplains. Despite being climate-risk areas for human habitation, they are often chosen due to their convenience and affordability for the poor. This justifies even the materials of the structures which in most cases are made of cardboard, mud and tin sheets, too substandard for human lives but forced to due to desperation for survival in cities (Tollin and Hamhaber

2017). All the structures are made without focus on the impact of climate change and this places human lives at risk when disaster strikes thus calling for concerted efforts.

The cooperation of the international world is important in the mitigation of climate change impacts. As highlighted by Stavins et al. (2014), the many climate change treaties and agreements being signed at a global level are relevant in bringing together nations to address climate issues. This is because climate change has become a common problem that is affecting everyone, be it the developing or the first world. One of the common factors is that built environments everywhere are known to create urban heat islands whose effects are to exacerbate heat waves which are hazardous to human lives (The Prairie Climate Centre 2017). However, though a common problem globally, Borquez et al. (2017) argue that for it to be effective, climate resilience should be applied with an understanding of the different contextual backgrounds as different contexts have influence on capacity and results on the ground. And, this brings in the notion of the relevance of various disciplines in the built environment and their contribution to climate resilience comprehensively and collaboratively when brought together.

3.2 Built Environment Disciplines and Climate Resilience

In the past two decades, tertiary education was unable to produce graduates that are well-equipped to face climate change and sustainability issues (Martin and Jucker 2005). In this era of climate change, educational curricula plays a key role in ensuring that learners and graduates are armed with a good knowledge base and skills that help even societies to meet responsively, sustainability and move it forward (Lewis et al. 2005). The need to foster sustainable development through education has been recognised by UNESCO (2004). This shows the relevance of disciplinary action in reclaiming the built environment through resilience. The Sustainable Development Commission (SDC) (2004) directly points out the need for universities to embed inter-disciplinary studies related to climate resilience to take up the challenge against climate change. However, one of the setbacks to resilient societies can be attributed to the lack of understanding of the resilience concept and sustainability as this often takes a one-sided environmental perspective (Lewis et al. 2005).

But, countering such mono-focused approach can be enhanced through promoting communication across disciplines and professions in the built environment and as well as among stakeholders. Spence et al. (2001) warns against the notion of disciplines operating in isolation of the other yet working towards countering the same challenge of climate change. The multiplicity of disciplines in the built environment makes it possible to counter climate change in a more comprehensive manner as the built environment itself is a complex system (Tougwa 2015). Ganah et al. (2008), highlights that a sustainable built environment's mandate is to enhance the quality of life of the people within the environment. This is imperative to foster climate resilience integration in the built environment disciplines such as construction through building-resilient and climate-sensitive infrastructures.

The ongoing rapid urban expansion and urban population increases together with the unprecedented climate change impacts are exacerbating the vulnerability of the poor to disaster especially in the global South. This is posing challenges to both the people and governments at large. Williams et al. (2019) argues that these areas already have informality, poverty, environmental and unemployment challenges and to add climate change on top increases the complexities. Thus, despite it being difficult to set an accurate timeline for the occurrence of climate change disasters, it is important to work on proactive actions that can equip communities to be resilient (Sato 2017).

The success of any development initiative is driven by funding for it to come through and this includes the drive towards attaining climate-resilient development. Resilience principles encompass a whole urban system as is illustrated by the framework and this means the need for well-channelling of funds towards successful overcoming of climate stressors and vulnerabilities. Efforts to push climate change through have been facilitated through the creation of the Green Climate Fund by the UNFCCC and the SDGs that targeted to raise at least USD$100 billion aimed at mitigating climate change impacts. The Paris COP21 of 2015 brought a new beginning for climate policies and finances to stir global action. This was supported by the Paris Agreement that encouraged the contribution of finance by developed countries towards climate mitigation in developing countries (Brazier 2015a). This makes Zimbabwe have more financial resource channels to build on climate change initiatives from the international world as finance is acknowledged as a crucial enabler to fighting climate change.

A win against climate change makes it easier to achieve other development targets as its manifestations have an effect upon communities (UNDESA 2016). Climate change impacts together with structural inequalities hold people in a vicious cycle of poverty that will take long to disentangle from. Resilience, among many other actioned strategies, goes a long way in fostering community awareness and preparedness to counter climate change impacts. A community's level of vulnerability is one of the major determining factors to the severity of the impacts (Singh et al. 2014). This means that, the greater the community's built-in resiliency, the greater are its chances to recover and cope with climatic disasters. Thus, major efforts need to be put in enhancing a community's resilience. Built environment disciplines, when well-crafted to suit contemporary climate change challenges, empower graduates to easily transition into cleaner jobs that place resilience and sustainability affront (Iyer-Raniga and Dalton 2017).

3.3 *Climate Change Vulnerabilities and the International World*

Climate change is one major challenge that equalises all nations be it the developing or the developed ones. However, the differences come on the severity of the impacts as the developed world seems to be more prepared to regain from the shocks. Bosher et al. (2007) highlights the experiences of the United Kingdom to flooding which, despite it being a fact that wetlands are not environmentally safe for housing, at least 27% of new housing in the country have been constructed on low-lying wetland areas thus placing them in a more vulnerable state to flooding. Such circumstances are also experienced in the developing world, though in particular, the settlements being unplanned and informal. However, in both cases, it still remains imperative to ensure climate-sensitive planning and development in order to protect humanity and development assets. According to Vijayan et al. (2019), the construction sector also plays a key role in initiating resilient housing and development. The United Kingdom, together with other European nations have made it a priority to foster innovative resilient solutions to construction. This makes it necessary to imbed resilient tenets into building and construction codes.

Built environment disciplines such as planning and construction need to begin to implement climate-sensitive designs. Lewis et al. (2005) argue that humans spend at least 80% of their lifetime in buildings and yet the buildings are failing to meet human needs. Approximately 50% of carbon emissions emanate from buildings in the United Kingdom as highlighted by Pett et al. (2004). A failure to adjust to climate change resilience expectations today means that many buildings in the future will be inhabitable. This makes it important to embed climate change principles within the built environment disciplines curricula to create an appropriate value set and knowledge (Sayce and Ellison 2003). Universities should take advantage of the multi-disciplinary professionals to collaborate towards building tangible climate resilience strategies as is advanced by the climate resilience framework. The framework makes it easier to address the issue of climate change from a holistic point of view (Mangizvo et al. 2015).

Also, Australia is one other country that has managed to foster climate resilience in its land use planning discipline which has successfully proven well in practice. According to Bajracharya et al. (2011), the country's local councils have been able to formulate climate-sensitive policies that guide infrastructural developments. China on the other hand, managed to strengthen and push against climate change impacts through enforcing early warning systems as well as effective tools for monitoring of disasters (Mazanhi et al. 2021). This equips communities to be ready when disaster strikes and thus saving lives through reinforced resilience. The developed world, due to the availability of sufficient finances, have been able to counter climate change and promote resilience. The success of climate-resilient policies varies contextually and what may work for country A may be detrimental for country B due to differing magnitudes and severity of the impacts. Therefore, it is important for policymakers to formulate context-based climate-resilient policies and one of the best guiding tools is the climate resilience framework that is a one-size-fits-all as it can be modified to suit a specific environment.

3.4 Climate Change Vulnerabilities and Resilience in Africa

Africa is also bearing the brunt of burden of climate change and is also participating in climate change policy formulations and action. The Africa Union Commission in 2015 set out Agenda 2063 which is targeted at transforming development in Africa through enhanced natural resources protection (Africa Union Commission 2015). This would involve the implementation of climate change adaptive measures for sustainable management of bio-diversity. After realising that the 2063 Agenda's success demands proactive efforts against climate change, the Climate Change and Resilient Development Strategy and Action Plan that runs from 2022 to 2032 was instituted (Africa Union 2022). It outlines the major parameters as well as key priorities towards improving the resilient capacities of African communities towards an enabling sustainable environment as driven by the global New Urban Agenda. This brings together collective efforts towards addressing climate change impacts and at the same time enhancing development integration. In order to foster sustainable development, resilient communities are a key component.

The challenges of shortage of housing in most urban centres and extreme poverty levels are driving a lot of people into informal settlements for survival. Most of these settlements are located in high-risk areas such as wetlands which are prone to climate change risks such as flooding. When already they are living under poor, vulnerable conditions, an occurrence of climate change disaster, as highlighted by USAID (2018), leaves them in far worse conditions than they were before. This leaves them trapped in poverty as each disaster would leave them in huge losses of life, assets and livelihoods disruption. Thus, building resilience becomes an urgent matter for the billions of people living in informal and highly vulnerable areas. Sustainable development promotes equality and equity, and in the midst of climate change disasters, resilience becomes the bridge to achieving that as communities need to be able to regain after experiencing disasters or rather become prepared for any uncertainties (Valencia et al. 2021). In the absence of climate change, already inequalities exist, and worsen in the presence of climate disasters, thus making it urgent to equip human settlements with resilience.

Despite being one of the most vulnerable continents, the sub-Saharan Africa region is said to be the least in contribution towards climate change action globally (Codjoe and Atiglo 2020). Despite the already existing development challenges in the region, climate change impacts are souring higher and more destructive due to the lack of economic and technological development in the region. This calls for urgent action and cooperation in the fight against climate change as it is a global phenomenon that hits hard despite a country's status though the developed countries manage the challenges better. Sub-Saharan Africa needs to invest in good governance that stirs citizenry participation. Climate change challenges have a contributory factor upon many other challenges and once the region manages to deal with its effects in a more resilient manner, it becomes easier to meet other challenges.

4 Methodology

Rapid and uncontrolled expansion of urban areas which has sprouted the growth in informal settlements has placed human lives and the built environment in danger of climate change and its impacts. The creation of many vulnerabilities that include the loss of lives, livelihoods and imminent poverty makes the issue of urban resilience of urgency. Over time, the worsening impacts of climate change are placing the built environment in a more vulnerable state that makes resilience of huge concern in managing sustainability. In order to proffer long lasting sustainable solutions to urban development, it is imperative to deepen the understanding of urban resilience through exploring how it can be applied to human settlements. The study thus makes use of a secondary research approach through exploring relevant and applicable journal articles, review of government climate change policies and strategies, case studies and remote sensing data that enable the understanding and applicability of climate resilience framework on the ground. This helps in providing precise and latest information relevant to the study.

5 Results

Zimbabwe is still a developing nation compassed with a load of challenges and development targets. With the imminence of climate change impacts, it is important to proactively ensure climate resilience strategies in both the current and future development initiatives rather than dealing with the challenges later on after disaster strikes. Sustainability and resilience principles in development may seem a daunting exercise, but its benefits carry weight in the long run (UNFCCC 2020). Financing climate action demands governance commitment that is primed at ensuring built environment resiliency. With the increasing urban population in the country and the constant rise in climate change disasters, channelling focus and resources on upgrading of human habitation is of necessity. The USAID (2018) denotes that upgrading of informal settlements, though likely to be costly is necessary to ensure climate resilience and human livelihood protection.

In response to global climate change, the international community has made efforts to develop climate response laws and strategies. In response to that, Zimbabwe formulated its National Climate Policy (2017) that is targeted at promoting an economy that is low-carbon as highlighted by Murombo et al. (2019). The policy is targeted at ensuring coordination and financing of climate resilience strategies. According to its provisions, the policy anchors on research, resource mobilisation and good governance among many other pillars to see the vision of a carbon-free

Zimbabwe successful. Enabling resilience is one of the priority matters of the policy that is necessitated with the carrying out of vulnerability assessments across sectors particularly the infrastructural sector. Despite the country's infrastructure being developed in view of climate change, the intensity of the impacts in the modern day demand more climate resilience emphasis.

While efforts are being done in the country through the shaping of policies, this does not remove the responsibility that the local citizens have on making climate resilience a success as they are directly at the interface where all the damages are taking place (Al-Humuisqani and AlGhamdi 2022). This is also supported by the COP20 that took place in 2014 in Lima with the participation of approximately 195 nations and 14,000 representatives (Gutierrez 2014). Conference of Parties are carried out annually and climate change and lowering of gaseous emissions are the major themes in the discussions. The COP expressed the need for active participation and collaboration of all parties its 'appeal for climate action.' Zimbabwe's representative bodies in the COP20 included the Farmers' Union and livelihoods programmes' bodies. Despite the efforts to participate in the COP, there remains a gap in the level of participation of the built environment disciplines. It can be argued that climate efforts in the country are biased towards the farming sector as it is the major livelihood and economic base for the country.

The country's climate endeavours are also guided by the National Climate Change Response Strategy that acknowledges climate change as a threat to humanity that calls for urgent action (Brazier 2015b). Achieving climate resilience in the country, though seemingly a daunting task, can be achieved with the cooperation and participation of citizens. Though Zimbabweans are already resilient and adaptable to some development challenges, they are acutely vulnerable to climate challenges (Jumbe et al. 2008). To instil urban resilience in the country, Al-Humuisqani and AlGhamdi (2022) suggest the need for resourcefulness, self-organisation and proper governance and these are also supported by the resilience framework highlighted earlier.

In view of Zimbabwe's experience, the framework hasn't been actively applied or rather it has been in play indirectly as the active role of disciplines curricula towards sustainability and the New Urban Agenda has been invisible. This is because much of the work on climate change resilience and adaptation in Zimbabwe is driven by various organisations and networks such as ZERO Zimbabwe, Climate Change Adaptation Africa (CACC), African Technology Policy System (ATPS) and CARE Zimbabwe among many other bodies (Dube and Chimbari 2009). Other key organisations as highlighted by Bandauko (2021) include the UNDP, Ministry of Local Government and UNICEF which hold a Partnership for Building Urban Resilience in Zimbabwe. The partnership goals are aimed at enhancing urban resilience and improving livelihoods for the unemployed and vulnerable groups of society in urban areas. Apparently, when it comes to climate action in the country, networks and organisations tend to play a bigger role more than academic institutions despite the eagerness by university students to explore climate change (Mangizvo et al. 2015). Due to that limited exposure, they tend to get climate knowledge from social media which in most cases is unreliable and distorted as lecturers may not be in a good position to impart climate knowledge (Cordero et al. 2008).

Built environment disciplines have a key responsibility in fostering sustainability through imbedding it in curricula due to their direct relation with property, land and construction (Lewis et al. 2005). Most of the built environment disciplines are aided by laws and regulations on the environment for example the spatial planning discipline. The University of Zimbabwe, among other educational institutions in the country managed to introduce the issue of climate change and sustainability through a review of some of the built environment disciplines that include spatial planning curriculum. This has brought in climate as a central aspect to planning education and training and planning institutions. The participation of the institution in urban resilience building in collaboration with development partners is an acknowledgement of the interconnectedness between the institution and communities at large. The students come from communities in which climate change takes a toll in most cases and by infusing climate knowledge in urban curricula, it becomes easy for climate information to cascade easily to communities. The infusion of climate change in the curricula helps improve climate literacy among students (Mangizvo et al. 2015).

Built environment disciplines, make it easier to map out vulnerabilities facing settlements and this is a good example of bridging the gap between curricula and practice. Geography and Information Science as a discipline, aids in mapping climate change hotspots and thus aid shaping problem-specific solutions towards climate resilience and the New Urban Agenda. A good example is that of Harare's informal settlements that follow river banks such as the Mukuvisi River. Some of the places include Paradise and Glen Norah C which are located close to Mukuvisi River where there are high chances of being affected by flooding. Another vulnerability risk that can be identified is related to the environment and the ecosystem. The environmental conditions in which these people live are far from being safe and make them more vulnerable to health hazards (Mavhima 2019). It is imperative in view of these risks that the government takes cognisance and makes effort to relocate the people to safer places. Mapping out such high-risk areas has shown how disciplinary action is taking part in climate resilience in the country.

Climate policies and laws have shown the urgency of the need to promote climate resilience and the need for cooperation among states as the issue of climate change is a global matter (Mazanhi et al. 2021). The slackness of one member to do their part in the fight against climate change has an indirect effect on those that are faithfully making efforts to fight against it. Thus, in a bid to join the forces, the country formulated the National Climate Policy of 2017 which works in support with the National Climate Change Response strategy. The aim of the policy is targeted towards creating a carbon-free and resilient Zimbabwe and lower vulnerabilities. The policy is aimed at achieving that vision through enhanced climate research, data collection and vulnerability assessments (Republic of Zimbabwe 2017). This fits well with the discipline of spatial planning as is taught at the University of Zimbabwe that fosters research and the carrying out of environmental impact assessments. There is need to have more of such national policies that integrate curricula and practice. Other supportive policies in the country include the Renewable Energy Policy, the Forestry Policy and the Drought Mitigation Policy among many other policies on environmental management and climate change mitigation (Brazier 2015b). These are also supported by the Environmental Management Act (Cha 20:24) that supports environmental protection (Mazanhi et al. 2021). Until climate change challenges have found long lasting solutions, there is no need to slack on the part of the government in fostering climate resilience (Kutywayo et al. 2013).

The Case of Cyclone Idai in the Eastern Highlands of Zimbabwe

The 2019 occurrence of Cyclone Idai disaster in Chimanimani and Chipinge Districts reminded of the importance of spatial planning in settlement development. The cyclone hit hard people who were settled in unsafe areas and conducted activities (such as cutting down of trees along steep slopes and rivers) that increased disaster risk. Many institutions including the Department of Civil Protection (DCP), Provincial and District Development Coordinators, Councils, the private sector and civil society were involved in responding to the disaster, but public institutions lacked technical, financial and logistical capacity and were more reactive rather than proactive (Chatiza 2019). Councils, local people and companies were the first responders and demonstrated good practice and innovation. However, the councils lacked capacity to provide services, particularly for resettling displaced urban residents who required serviced land and assistance in house construction. Capacity gaps were evident at all levels of governance, for instance, the National Army lacked equipment (helicopters) that could handle extreme weather conditions including mist, strong winds and continuous heavy rain. Cyclone Idai exposed ineffective national early warning systems, knowledge building and uncoordinated evacuation plans (Chatiza 2019).

6 Discussion

More than 50% world's population resides in urban areas and the cities are already facing the brunt of urban challenges making them more vulnerable to climate change disasters. Shamout et al. (2022) highlight the need to increase the level of resilience in the urban areas especially in view of the many global challenges that are emanating especially climate change. Resilience has become the common topic in sustainability in both national and international debates (Da Silva and Morera 2014). The Sustainable Development Goals have explicitly provided for resilience in goals number 9 that targets the development of resilient infrastructure and number 11 that aims at making human settlements and cities safe, resilient, sustainable and inclusive (United Nations General Assembly 2015). The awakening to the importance of resilience in the built environment disciplines was adopted late when compared with other climate-related disciplines (Meerow and Newell 2019). However, despite that, resilience is quickly gaining ground and popularity in contemporary debates, policymaking and development as it has proven to be a tool for sustainable planning and development.

Most of the climate change actions implemented in the Zimbabwe have been aiming at reducing climate vulnerabilities of human settlements and improving their resilience and adaptation to climate hazards. This is also presented by the climate resilience framework which makes resilience the focal point through assessment of climate vulnerabilities (Cordoso et al. 2020). Successful climate-sensitive development cannot take place in bits and pieces but requires a procedural framework that applies from the development initiation stage to its implementation and monitoring. The climate resilience framework proves to be a comprehensive tool that the country needs to integrate in its climate actions. It will foster accountability and much needed guidelines on reducing vulnerabilities (USAID 2014).

The contribution of the built environment disciplines in Zimbabwe's universities is still slow to catch up with the rapid climate change. In a more specific sense, the built environment encompasses all aspects that are humanly created to meet human needs and comfort (Haigh and Amaratunga 2010). This means that a damage to the built environment by climate change hazards will directly impact human welfare. Most economic disasters leave human societies in a more crippled state where they are severely impaired to recover from and this often results in a multiplicity of challenges that exacerbate the vulnerabilities of people and thus the emergence of poverty. An emphasis on the built environment disciplines in local institutions is an acknowledgement of its capacity to provide deep knowledge and capacity through which humans can lower climate hazard risk and thus be prepared when a disaster occurs. Vulnerability to climate change in post-disaster crisis is worsened by loss of infrastructure and buildings and without much responsive action, it makes the community even more vulnerable to disasters that would follow thereafter.

Being a complex system and due to the local, individual nature and context portrayed by the built environment, applying generic sustainable solutions as presented in the New Urban Agenda and the Sustainable Development Goals (SDGs), is restricted (Haigh and Amaratunga 2010). Therefore, this makes it desirable to involve the built environment disciplines through an inter-disciplinary action that targets the enhancement of climate resilience in an urban setup. An inter-disciplinary action takes place when the built environment disciplines offer their definitive goals and conceptual understanding of climate change towards a collective action. A successful inter-disciplinary approach, especially in the light of creating sustainably resilient communities, would be able to involve and value societies and their identities rather than sole reliance on singular disciplinary identities. Since the academic disciplines are formulated to empower and equip students and society on good climate resilience measures, it would be imperative to offer an equal platform for co-participation with the community members and knowledge exchange. Though Zimbabwe's population is considered resilient, this only applies to economic and other social matters but when it comes to climate issues, urgent action needs to be done to raise the resilient standards from the vulnerabilities the populations are faced with.

The success of built environment in driving climate resilience and sustainability as supported by the New Urban Agenda requires good governance. Teaching modules and classes that are climate-related is not in itself sufficient to enable the achievement of climate resilience. This is supported by Charles and Chang-Richards (2022) who denote that public institutions and officials ought to exercise authority in fostering good policymaking to support climate resilience initiatives. Poor governance and lack of political will fall among the major contributors to policy failure and thus need to be enhanced so that the curricula efforts of the built environment disciplines do not go in vain (Max Lock Centre 2009). Government-led resilience initiatives are important when implementing and this will aid in bridging the gap between built environment theory and practice.

The case of Cyclone Idai demonstrates vulnerability at all levels of governance. The protracted political and economic crisis has resulted in increased poverty and urban informality. Residents of informal settlements experience multiple deprivations such as unemployment, poverty, insecurity and inequality in access to infrastructure (Matamanda and Mphambukeli 2022). The multiple deprivation increases their vulnerability to extreme weather events such as floods. A critical analysis of disasters in the African context reveals multi-risk, multi-vulnerability and multi-hazard (Cotti et al. 2022). Multi-risk analysis looks into the presence of multiple risks and related potential losses. Multi-vulnerability is a measure of the different vulnerabilities of multiple exposed elements. Multi-hazard refers to multiple hazards in the sense of hazard interactions and cascades. Spatial analysis for hazard exposure components, construction of a composite indicator for the vulnerability component and matrix-based aggregation for risk are recommended methodologies for risk assessment towards multi-risk (Cotti et al. 2022).

In steering the New Urban Agenda, the built environment in informal settlements amplifies the vulnerability of individuals and groups (Birch 2021; Sait 2021). Zimbabwean slum communities are generally weakened through exposure to poor governance, economic stress and natural disasters such as floods and drought. Incoherent urban policies, economic crisis, infrastructure collapse or its inexistence in some informal settlements have been the order of the day. Urban policies have been incoherent where development has been permitted in the absence of requisite infrastructure, responsibility for revenue collection has been centralised in state enterprises such as ZIMRA, ZINARA and ZESA; and enforcement of regulations has been weak. In this vulnerability context, the role of the built environment disciplines and practice towards the New Urban Agenda becomes a mammoth task.

Most of the climate change disasters that take place, although they are termed emergency disasters, will in most cases be expected (Bosher et al. 2007). The failure to be proactive in planning for resilience worsens the severity of the impacts. With the application of the climate resilience framework in development, it becomes easier to develop climate-sensitive human settlements that are sustainable. In order to attain resilient settlements with well-equipped citizens, it is important to be proactive in planning and implementation of climate strategies in a more systematic way. This may call for resettlement of people in environmentally sensitive areas and even a change in design structures. Building codes should also be reviewed in order to incorporate climate matters in the process. However, the many challenges already facing Zimbabwe's cities make them incapacitated for climate action. But, the climate resilience framework, though time-consuming to put into action (Home 2021), can be implemented with minimal resources thus leaving no reason for reluctance on the part of developers.

7 Conclusion

The chapter has explored the relevance of climate change resilience in urban settlements and the built environment at large. The climate resilience framework as suggested in the chapter offers a comprehensive platform for developers and decision-makers to proffer climate-sensitive solutions to the built environment while at the same time equipping the citizens with readiness to counter climate challenges. Extreme climate change impacts have proven detrimental to the existing built environment. In view of this, it is imperative to factor climate resilience in planning as a means to capacitate communities in the long run, in order to attain the global concern on sustainable development and the New Urban Agenda. The framework facilitates planning for adaptation to climate change in cities as it goes beyond the focus on impacts but interrogates the existing vulnerabilities and how specific actions can be initiated. The study recommends the use of the framework in order to ensure the integration of infrastructure, social, institutional and ecological resilience factors. It simplifies development through its organisation of a rubric that fits well in both theory and practice.

Climate resilience is a crucial factor in development and settlement management in urban areas. The study sought to bridge the gap between the built environment disciplines' contribution to climate change mitigation and adaptation practices through the novelty of the resilience framework's applicability. A realisation that countering climate change in bits and pieces is insufficient to comprehensively cover climate issues in an urban setup has made the climate resilience framework a great tool both in theory and practice. The research contributes to sustainable and resilient-built environments in both the inter-disciplinary and post-disciplinary perspectives. Human settlements cannot be separated from climate change and hence the need to emphasise on resilience planning in order to create future, climate-smart built environments. With the intensity increase in climate change impacts, it is long overdue for the Zimbabwe to engage in resiliency. The built environment disciplines play a key role in influencing climate resilience and the framework presented in the chapter provides a comprehensive notion of how various disciplines ought to work hand in glove to achieve the common goal.

References

African Union (2022) Africa's Climate Change and Resilient Development Strategy and Action Plan (2022–2032), Addis Ababa, African Union Commission

African Union Commission (2015) Agenda 2063: The Africa We Want, Addis Ababa, African Union Commission.

Al-Humuisqani MM, AlGhamdi S (2022) The built environment resilience qualities to climate change impact: concepts, frameworks and directions for future research. Sustain Cities Soc 80:1–17

Bajracharya B, Childs I, Hastings P (2011) Role of local government in disaster management: findings from regional towns in Queensland. https://www.semanticscholar.org/paper/Role-of-local-government-in-disaster-management%3A-in-BajracharyaChilds/b36f22693006beaf34b3f4b101092f77406ed50a

Bandauko E (2021) Building urban resilience in the post-2015 development agenda: a case study of Harare, Zimbabwe. In: Chirisa I, Chigudu A (eds) Resilience and sustainability in urban Africa. Advances in 21st century human settlements. Springer, Singapore. https://doi.org/10.1007/978-981-16-3288-4_6

Beringer AL, Inmuong Y, Kaomuangnoi K (2019) Understanding urban vulnerabilities to climate change impacts in Khon Kaen and Mukdahan in Thailand. Soc Sci Asia 4(4):1–13

Birch H (2021) Urban governance and disease outbreaks: Cholera in Harare and Ebola in Monrovia. In: Home R (ed) Land issues for urban governance in Sub-Saharan Africa. Springer, Cham, Switzerland, pp 299–315

Borquez R, Aldunce P, Adler C (2017) Resilience to climate change: from theory to practice through co-production of knowledge in Chile. Sustain Sci 12:163–176

Bosher L, Carrilo P, Dainty A, Glass J, Price A (2007) Realising a resilient and sustainable built environment: towards a strategic agenda for the United Kingdom. Disasters 31(3):236–255

Brand F, Jax K (2007) Focusing the meaning(s) of resilience: resilience as a descriptive concept and a boundary object. Ecol Soc 12(1):23

Brazier A (2015a) Climate change in Zimbabwe: facts for planners and decision makers. Konrad Adenauer Stiftung

Brazier A (2015b) Climate change in Zimbabwe: a guide for planners and decision makers. Konrad Adenauer Stiftung

Charles CH and Chang-Richards A (2022) Providing a framework for post-disaster resilience factors in buildings and infrastructure from end-users perspectives: case study in Caribbean Island States. Int J Disaster Resilience Built Environ, Emerald Publishing Limited. Accessed from https://doi.org/10.1108/IJDRBE-02-2021-0020 31/01/2023

Chatiza K (2019) Cyclone Idai in Zimbabwe: an analysis of the policy implications for post-disaster institutional development to strengthen disaster risk management, Oxfam, Harare, Zimbabwe

Chynoweth P (2006) Legal scholarship in the built environment: epistemological considerations and cultural challenges. In: The construction and research conference of the Royal Institution of the Chartered Surveyors, RICS

Codjoe SNA, Atiglo DY (2020) The implications of extreme weather events for attaining the sustainable development goals in Sub-Saharan Africa. Front Clim 2, Article 592658

Cordero EC, Todd AM, Abellera D (2008) Climate change education and the ecological footprint. Bull Am Meteor Soc 89(6):865–872

Cordoso MA, Brito RS, Almeida MC (2020) Approach to develop a climate change resilience assessment framework. H2Open J 3(1):77–88

Cotti D, Harb M, Hadri A, Aboufirass M, Chaham KR, Libertino A, Campo L, Trasforini E, Kratzschmar E, Bellert F, Hagenlocher M (2022) An integrated multi-risk assessment for floods and drought in the Marrakech-Safi Region (Morocco). Front Water 4:886648

Da Silva J, Morera B (2014) City resilience framework. Arup & Rockefeller Foundation, London/New York. Available at http://publications.arup.com/Publications/C/City_Resilience_Framework.aspx

Dube OP, Chimbari MJ (2009) Documentation of climate change and human health in southern Africa. University of Copenhagen, Copenhagen, Denmark

Ganah A, Pye A, Hall G (2008) The role of knowledge transfer in sustainability research in the built environment discipline. In: Dainty A (ed) Proceedings of the 24th annual ARCOM conference, 1–3 September 2008, Cardiff, UK, Association of Researchers in Construction Management, pp 299–307

Gutierrez WB (2014) Climate change conference: results of the COP20 in Peru. Sci Soc Notes 4(2):142–143. Accessed from https://doi.org/10.18259/acs.2014015. 31/01/2023

Haigh R, Amaratunga D (2010) An integrative review of the built environment discipline's role in the development of society's resilience to disasters. Int J Disaster Resilience Built Environ 1(1):11–24

Hassler U, Kohler N (2014) Resilience in the built environment. Build Res Inf 42(2):119–129

Home R (2021) Urban law and resilience challenges of climate change in MENA region. In: Urban and resilience challenges, pp 153–168

Iyer-Raniga W, Dalton T (2017) Challenges in aligning the architecture profession in Indonesia for climate change and sustainability. Procedia Eng 180:1733–1743

Jumbe CBL, Wiyo KA, Njema E, Msiska FBM (2008) The role of government, donors, civil society and the private sector in climate change adaptation in Malawi: scoping study. Discussion Paper, Christian Aid, Innovative Fund Africa, Lilongwe, Malawi

Kutywayo D, Chemura A, Kusena W, Chidoko P and Mahoya C (2013) The impact of climate change on the potential distribution of agricultural pests: the case of the coffee White Stem Borer (Monochamus leuconotusP.). PLoS One 8(8):e73432

Lewis A, Sayce S, Ellison L (2005) Education for sustainable development in the built environment disciplines. Centre for Education in the Built Environment, CEBE, Working Paper No. 9. Cardiff University, Cardiff, UK

Mangizvo RV, Chirume S, Mahlatini PJ (2015) Limited awareness on climate change at Zimbabwe Open University (Midlands Region): a Deterrent on information dissemination. Int J Humanit Soc Sci Educ (IJHSSE) 2(5):188–198

Martin S, Jucker R (2005) Educating Earth literate leaders. J Geogr High Educ 29(1):19–29

Matamanda AR, Mphambukeli TN, Chirisa I (2022) Exploring water-gender-health nexus in human settlements in Hopley, Harare. Cities Health 8(1):44–58

Mavhima B (2019) Building urban resilience in the face of climate change in Harare. J Urban Syst Innov Resilience Zimbabwe 1(1 & 2):149–169

Max Lock Centre (2009) The built environment professions in disaster risk reduction and response: a guide for humanitarian agencies. MLC Press, University of Westminster

Mazanhi P, Nyamugadza E and Chirisa I (2021) Climate change and human settlements: towards a strategic framework for sustainable and resilient land use planning and practice in Zimbabwe. In: Nyikahadzoi K, Mhlanga L (eds) Climate change impact, adaptation and mitigation in Zimbabwe: case studies from Zimbabwe's urban and rural areas. Konrad Adenauer Stiftung, pp 135–158

Meerow S, Newell JP (2019) Urban resilience for whom, what, when, where and why? Urban Geogr 40(3):309–329

Murombo T, Dhliwayo M, Dhlakama T (2019) Climate change law in Zimbabwe: concepts and insights. Konrad Adenauer Foundation, Harare

Opitz-Stapleton S, Guibert G, MacClune K, MacClune K, Sabbag L, Tyler S (2010) Only death is certain, yet you still get out of bed in the morning: or observations on the use of climate information in adaptation and resilience practice. Climate Resilience in Concept and Practice Series Working Paper No. 2. Institute for Social and Environmental Transition (ISET), Boulder, CO

Pett J, Guertler P, Hugh M, Kaplan Z, Smith W (2004) Asset value implications of low energy offices phase 2 report. Association for the Conservation of Energy. United Kingdom, London

Republic of Zimbabwe (2017) National climate policy. Ministry of Environment, Water and Climate, Harare, Government Printers

Sait MS (2021) Should Monrovian communities agree to voluntary slum relocations: land, gender and urban governance. In: Home R (ed) Land issues for urban governance in Sub-Saharan Africa. Springer, Cham, Switzerland, pp 339–354

Sato S (2017) Climate change, the built environment and triple-helix innovation. Energy Procedia 143:843–850

Satterthwaitte D, Archer D, Colenbrander S, Dodman, Hardoy J, Mitlin D, Patel S (2020) Building resilience to climate change in informal settlements. One Earth Rev

Sayce S, Ellison L (2003) Integrating sustainability into the appraisal of property worth: identifying appropriate indicators of sustainability. In: The American Real Estate and Urban Economics Association conference, Skye, Scotland

Shamout S, Boarin P, Mannakkara S (2022) Understanding resilience in the built environment: going beyond disaster mitigation. In: The 54th international conference of the Architectural Science Association (ANZAScA) November 2020, Auckland University of Technology, Auckland

Singh SR, Eghdami MR, Singh S (2014) The concept of social vulnerability: a review from disaster perspectives. Int J Interdisc Multi Stud (IJMS) 1(6):71–82

Spence R, Macmillan S, Kirby P (2001) Interdisciplinary design and practice. Thomas Telford, London

Stavins R, Zou J, Brewer T, Conte Grand M, den Elzen M, Finus M, Gupta J, Höhne N, Lee MK, Michaelowa A, Paterson M, Ramakrishna K, Wen G, Wiener J, Winkler H (2014) International cooperation: agreements and instruments. In: Edenhofer O, Pichs-Madruga R, Sokona Y, Farahani E, Kadner S, Seyboth K, Adler A, Baum I, Brunner S, Eickemeier P, Kriemann B, Savolainen J, Schlömer S, von Stechow C, Zwickel T, Minx JC (eds) Climate change 2014: mitigation of climate change. Contribution of working group III to the fifth assessment report of the intergovernmental panel on climate change. Cambridge University Press, Cambridge, United Kingdom and New York, NY, USA, pp 1001–1082

Sustainable Development Commission (2004) Shows promise but must try harder. https://www.bing.com/ck/a. Accessed 31/01/2023

The Prairie Climate Centre (2017) Building a climate resilient city: the built environment. The Prairie Climate Centre

Thondhlana G, Mubaya CP, McClure A, Amaka-Otchere ABK (2021) Facilitating urban sustainability through transdisciplinary (TD) research: lessons from Ghana, South Africa and Zimbabwe. Sustainability 13:6205

Tollin N, Hamhaber J (2017) Sustainable resilient cities: sustainable development goals, new urban agenda and the Paris Agreement. Energy, Ambience Innov, 8–15

Tougwa FN (2015) The built environment discipline and its research at doctoral level. University of Salford, Manchester, United Kingdom

Tyler S, Moench M (2012) A framework for urban climate resilience. Clim Dev 4(4):311–326

UNDESA (2016) Climate change resilience: an opportunity for reducing inequalities. World Economic Survey 2016, UNDESA

United Nations Educational, Scientific and Cultural Organization (UNESCO) (2004) United Nations decade of education for sustainable development 2005–2014 implementation scheme. UNESCO, Bonn, Germany

UNFCCC (2020) How to COP, a handbook for hosting United Nations Climate Change Conferences. UNFCCC, Bonn, Germany

UNFPA (2018) The power of choice: reproductive rights and the demographic transition. The State of the World Population 2018, UNFPA, New York, USA

UNGA (2015) Transforming our world: the 2030 agenda for sustainable development. United Nations General Assembly (UNGA), New York, USA.

UNHABITAT (2014) 2014 Revision of the world urbanization prospects. UNHABITAT, New York, USA

UNHABITAT (2016) The new urban agenda. UNHABITAT, Quito, Ecuador

UNHABITAT (2019) Guidelines for reporting on the implementation of the new urban agenda. UNHABITAT, Nairobi, Kenya

USAID (2014) Climate-resilient development: a framework for understanding and addressing climate change. USAID

Valencia SSC, Simon D, Croese S, Davison A, Diprose K, Srinivas AK, Nesprias, Nordquist M, Oloko T, Sharma T, Buck T, Versace I (2021) Urban climate resilience and its link to global sustainability agendas. The Palgrave Handbook of Climate Resilient Societies, Accessed on 31/01/2023 on https://link.springer.com/referenceworkentry/10.1007/978-3-030-32811-5_131-2

Vijayan V, Raj AS, Varghese MM, Panicker SL, Mohan D (2019) A study on sustainable and cost-effective building construction in housing sector. J Emerg Technol Innov Res (JETR) 6(6):563–571

Williams DS, Sutherland CG, Scheffran J, Costa MM, Celliers L (2019) Vulnerability of informal settlements in the context of rapid urbanization and climate change. Environ Urban 31(1):1–10. https://doi.org/10.1177/0956247818819694

The New Urban Agenda in Zimbabwe: Policy and the Future

Charles Chavunduka, Innocent Chirisa, and Edith Risinamhodzi

Abstract This final chapter synthesises the initiatives and issues identified in the different disciplines and their practice in the built environment and how they have been steering the New Urban Agenda in Zimbabwe in the context of a changing global environment. The initiatives include those that harness smart technologies, academia thrust on innovation and business linkages as well as those that have looked into shared value creation. Identified issues related to gaps in built environment curriculum and practices, and constraining legal and institutional frameworks. The chapter draws policy implications for more effective contribution of built environment disciplines to sustainable development. A New Urban Agenda implementation plan is identified as a critical need for guiding actors. Further, the chapter explores the future of the New Urban Agenda in Zimbabwe over the remainder of the 2020s and beyond. Improving the rate of progress towards the New Urban Agenda will require shaping the built environment sciences and practice for delivery. This will involve revamping curricula, business re-engineering and professionalisation of all built environment sciences.

Keywords New Urban Agenda · Built environment · Curriculum · Smart technologies · Professionalism

---

C. Chavunduka (✉)
Faculty of Engineering and the Built Environment, University of Zimbabwe, Box MP 167, Mt. Pleasant, Harare, Zimbabwe
e-mail: cchavunduka@eng.uz.ac.zw

I. Chirisa
Zimbabwe Ezekiel Guti University, Bindura, Zimbabwe

Department of Urban and Regional Planning, University of the Free State, Bloemfontein, South Africa

E. Risinamhodzi
Department of Architecture and Real Estate, University of Zimbabwe, Box MP 167, Mt. Pleasant, Harare, Zimbabwe

C. Chavunduka and I. Chirisa (eds.), *New Urban Agenda in Zimbabwe*,
Advances in 21st Century Human Settlements,
https://doi.org/10.1007/978-981-97-3199-2_13

1 Introduction

In this concluding chapter we hone in on some of the key areas emerging from an extensive exploration of the role of built environment disciplines in steering the NUA. Synthesising key messages from the chapters in a uniform way is ambitious nor would it do justice to the nuances which currently exist.

Globally and regionally, there are networks that exist to support the creation of a sustainable built environment. These networks have the potential to advance sustainable development goals and scale-up transformative change. The networks have been effective mechanisms in raising awareness and taking action related to environmental protection, and more recently, on issues related to sustainability, such as climate change, biodiversity, water, poverty alleviation, sustainable consumption and production. In the area of climate change, for example, there has been increasing demand for proactive building and infrastructure approaches that make use of more integrated land use-transport planning.

Some of the key networks include the Global Building Information Modelling (BIM) Network, One Planet Network, Strategy of Education for Sustainable Development in Sub-Saharan Africa and Sustainable Development Solutions Network. The Global BIM Network connects international public sector representatives and multilateral organisations with the aim of advancing the digitalisation of the global built environment and sharing the resulting benefits. The digital platform aims at facilitating the sharing of knowledge on building information model processes, enabling international collaboration and raising awareness of innovative research and technology developments. The One Planet Network was adopted in 2012 at the World Summit on Sustainable Development to implement the 10-Year Framework of Programmes on Sustainable Consumption and Production (SCP), which supports the global shift to SCP and the achievement of SDG 12. It brings together actors from all regions and sectors and harnesses their expertise, resources, innovation and commitment towards more sustainable production and consumption patterns (Iyer-Raniga and Huovila 2020). The Strategy of Education and Sustainable Development for Sub-Saharan Africa was born from the need for education to address the growing environmental challenges facing the planet (Leicht et al. 2018). In order to do this, education must change to provide the knowledge, skills, values and attitudes that empower learners to contribute to sustainable development. The UN Sustainable Development Solutions Network (SDSN) mobilises global scientific and technological expertise to promote practical solutions for sustainable development including the implementation of the SDGs and Paris Climate Agreement. SDSN works closely with United Nations agencies, multilateral financing institutions, the private sector and civil society.

Shaping the built environment sciences and practice remains a priority for African countries in order to achieve the NUA. This is so because economic growth potential lies in rapidly urbanising cities and the built environment provides critical infrastructure for the urban economy (Roxburgh et al. 2011; MacLennan and Stacey 2016). The findings in this book show explosive urbanisation rates in Zimbabwe. However, this is taking place against a backdrop of declining industrial base and inadequate infrastructural expansion. The combination of rapid urbanisation, a declining industrial base and low investment in infrastructure expansion has resulted in more people being absorbed in the informal economy (Chavunduka 2020; Chavunduka and Chaonwa-Gaza 2021). Rising informality and spatialised inequality have been the hallmark of the urbanisation of poverty (MacLennan and Stacey 2016). Mashingaidze and Kanonhuhwa described how the urban public transport sector has grown informal in the context of protracted economic crisis.

Increased urban informality has heightened concerns about the environment and climate change which is threatening rural-based agricultural livelihoods, thus propelling rural–urban migration. Natural disasters are expected to become more frequent, and this raises the critical role of other domains such as natural disaster and risk management in the built environment. The destruction of infrastructure in eastern Zimbabwe by Cyclone Idai in 2019 indicates the growing need for resilient built environments. Geographic information systems (GIS) and remote sensing could draw on technological innovations to develop better early warning systems and map out natural disaster-prone areas. In the context of climate change, the biophysical environment is a key component in shaping the built environment because it presents multiple opportunities and resources that are critical to the functioning of cities. The biophysical environmental aspects considered are water, geomorphological assets, climate, soil and ecological resources.

With the built environment having a significant global footprint in terms of natural resource consumption, air pollution and waste generation, it is recognised as one of the areas that require immediate intervention. The chapter by Shumba highlighted the need for environmentally sensitive built environment initiatives. A case is made for the construction industry to embrace the green concept which enables more economical and environmentally friendly construction practices. It is a call for an ecological and economical city. An ecological and economical city builds on the synergy and interdependence of ecological and economic sustainability, and their fundamental ability to reinforce each other in the urban context (Bibri 2020). It allows citizens to live and work using minimum resources through the sustainable use of energy, water and transport systems among other resources. This concept has great potential of creating sustainable and resilient cities in developing countries if integrated into the development of new settlements or redevelopment of existing ones.

A binding constraint to the implementation of the NUA has been a weak state which is characterised by incoherent policy and institutional frameworks, outdated and rigid legal frameworks, top-down approaches to planning and rampant corruption. A coherent and up-to-date spatial planning policy, legal and institutional framework contribute to a sustainable built environment. Spatial planning legislation and model building by-laws were passed in 1976 and 1977, respectively, but they have not accommodated changes over time such as those relating to the development of new building technologies. Top-down planning and budgeting has limited local autonomy and the scope for adoption of collaborative approaches (UN-Habitat 2009;

Sarkissian 2005; Watson 2009). Could devolution of power and authority to lower-level government structures such as urban and rural district councils provide the answer? Mazongonda and Chigudu have shown that efforts at devolution of power to lower-level tiers of government have been slow and limited with financial control still retained by central government. This has reduced the capacity of local authorities to effectively carry out objectives of the NUA.

The next sections discuss the emerging themes and ideas, which the preceding chapters have developed. Shaping the built environment sciences and practice for steering the NUA in Zimbabwe will need to be embedded in built environment education and practice, the adoption of new technologies and coherent policy framework.

2 Built Environment Education

Built environment education comprises various sciences which provide a knowledge base for urban development. There has been concern about inadequate theory of the built environment, on which methods and practices can be based (Koskela 2008). Absent has been the glue that holds built environment disciplines together. Theory building in the inter-discipline has been fragmented, under-resourced and explored from limited perspectives of individual disciplines. Indeed, the built environment inter-discipline is an abstract notion and attempts at developing it into substantive knowledge that has been facing resistance (Koskela 2008). The preceding chapters recounted a multidisciplinary rather than interdisciplinary approach to understanding of the built environment. Questions are still posed concerning the targeted epistemic and practical gains from a built environment thought.

With the idea of a unified theory of the built environment rejected, Cairns (2008) suggests an approach that endeavours to enable all relevant and applicable theories and concepts to be brought to bear on the problems of the built environment. The approach draws upon recent interpretations of Aristotle's concept of phronesis, or practical wisdom. This pragmatic approach has been associated with applied science and technology more than with pure science (Kitchener 1999). These ideas conceptualise a theory of the built environment as something to strive for but bridged by a holistic approach to the inter-discipline.

In order to solve issues of sustainable development there is need to adopt a holistic and interdisciplinary approach which allows collaboration among built environment sciences. Some disciplines, for example spatial planning and architecture, have developed aspects of the built environment approach. The spatial planning chapter described the evolution of the built environment focus when the University of Zimbabwe's undergraduate degree in rural and urban planning study programme incorporated key built environment inputs like economics, building technology and construction economics in the training. This opened the door for real estate, valuation, quantity surveying, construction project management and urban design; a development which impacted on practice by greatly diversifying options for planning graduates. Opportunities exist for furthering collaboration and forging synergies with land surveying and engineering sciences. Already disciplinary boundaries are softening as has been the case between spatial planning and real estate. These developments will create more competitive and versatile graduates. The interdisciplinary approach is more in tune with technological and project requirements for data sharing and multi-skill teamwork.

3 Built Environment Practice

The built environment profession suffers from industry fragmentation, intense competition, limited investment in research and development and new challenges including IT advancements, increasing public expectations in environmental protection and enhancement (Kiviniemi 2009). In the Zimbabwean context, professional institutes are organised along traditional disciplinary boundaries. Built environment professional institutes operate in silos and there is no collaboration among them. The institutes focus on protecting their individual professions and creating a name for themselves within the built environment. Built environment professionals work independent of each other which undermines the role of productive collaboration and limits effective implementation of the NUA.

Interdisciplinary work among professionals enables effective implementation of the NUA in that it creates a platform for tackling complex built environment challenges which often transcend disciplinary boundaries. However, the various extents to which different built environment institutes in Zimbabwe have developed complicate the possible ways in which they can engage. Some of the professions draw from different philosophical and methodological traditions which further complicate the possibility of collaboration and coordination. Nonetheless, there is need for an institutional arrangement that will allow collaboration and coordination of built environment professions. This can take the form of a Council of Built Environment Professions—as an apex body for architects, engineers, planners, realtors, land surveyors, lighting designers and landscape architects. The Council would be charged with the responsibility to manage the professions, initiate research and collaboration. For instance, the Council of Built Environment Professions Australia (BEPA) worked with the (then) Department of Industry, Tourism and Resources to deliver an Action Agenda for this important industry sector (Kiviniemi 2009). The proposed Council could be partly funded through subscriptions by the institutes and partly from its own revenue raising activities.

4 The Built Environment Future

It is anticipated that urban informality will become a key feature of the built environment in Zimbabwe and other developing countries. Pervasive informal growth will present new challenges and opportunities within the built environment. With the right to the city, social justice and inclusivity discourse, there is growing impetus to rethink the ways in which built environment professionals have intervened in urban informal settlements.

Technology is expected to have revolutionary impacts on ways in which urban informality is managed. It has the power to reduce informality and increase security of tenure in the built environment. The Government of Zimbabwe embarked on upgrading slums in urban settings and drone-based aerial survey has offered opportunity for practitioners to accurately capture topographic base map together with an orthophoto, digital elevation and surface models which have aided the planning and regularisation process. Looking into the future, community-based small-scale technological interventions are set to increase. In addition to government efforts, it is important for built environment professionals to draw on these as they harness context-specific knowledge and offer promise for sustainability.

While the role of government in trunk infrastructure development is recognised, there is growing impetus to draw on stakeholder solutions. Traditional approaches relied on large-scale centralised networked infrastructural interventions but there is growing confidence in experimenting with community-based, home grown decentralised infrastructural configurations. These developments have been widening the number of stakeholders in the built environment and presenting new opportunities in governance. While the built environment was traditionally the domain of engineers and architects, international and local trends indicate a wide range of interests vested in the built environment. From residents, civil society, environmentalists and multilateral international organisations, the built environment is increasingly mediating a wide range of actors with often divergent and conflicting interests. With increasing technological advancement and its role in allowing access to information, residents are increasingly going to have a greater say in how the built environment is made, used and governed. This entails that the future of the built environment will involve experimenting with new governance models. These should be participatory and allow multistakeholder engagements. Considering these changes, sustainable practice in the built environment should combine the hard skills of engineering and design with the soft skills of negotiation and facilitation.

The findings from all chapters converge on the impact of globalisation, information and communication technologies in transforming the built environment and into the future (Roxburgh et al. 2011). Industry innovation is addressing the challenge of ever-increasing population in a resource-constrained world. Transforming buildings and infrastructure to become more sustainable elements of the built environment is a key challenge of the sciences as well as government at all levels. At the heart of the transformation is the promise of virtual building, an ability to assess the performance of a proposed built asset (e.g., lifetime cost, environmental impact, social benefit and locality impact) prior to construction (Newton et al. 2009). For example, the use of 4D CAD enables virtual construction before building in reality, enables other emerging processes such as virtual project teams and the off-site manufacturing of buildings. Central to virtual building is the building information model (BIM), and integrative digital technology that allows information-sharing among disciplines. It allows the design of a city not to be confined to an individual but a multidisciplinary team in order to minimise errors and omissions (Kiviniemi 2009).

BIM allows all stakeholders to work concurrently on a project in a modelled representation of a complex system, which is the built environment. It presents opportunities for built environment professionals to ride on technological transitions and produce smart developments that are relevant and attractive for investment. At the same time, it also presents unique challenges that stretch the current knowledge and practice within the built environment disciplines. It will transform how professionals within the built environment plan present and implement their work packages. This demands continuous capacity building among built environment professionals so that they are adequately equipped to respond to these challenges.

5 Political Will

A significant amount of literature identifies the real source of dysfunction and incoherence in urban policy across Africa—but especially in Zimbabwe—as the absence of political will rather than lack of capacity by various governmental organisations (Hanson et al. 2012; Poncian and Mgaya 2015; Unya et al. 2022). There has been a severe scarcity of committed leadership to identify problems, and begin to rectify them. Increasingly, much of Africa's problems have been ascribed to leadership and governance weaknesses. This realisation led African leaders to come up with a new economic plan in the form of the New Partnership for African Development (NEPAD) and the African Peer Review Mechanism, but benefits from the new institutions are yet to be realised (Unya et al. 2022). Addressing the leadership deficit—and related governance weaknesses—is a prerequisite for sustainable development. In Zimbabwe, rural and urban poverty has been on the rise in the context of macro-economic instability and currency crisis. The preceding chapters have shown how poor policies, corruption, clientelism and the politics of patronage have been impediments to sustainable and equitable development.

6 Conclusion and Implications for Policy

Various global agendas, for example, Local Agenda 21 and the Millennium Development Goals (MDGs) were agreed upon to address social, economic and environmental issues relating to development with varying degrees of impact, particularly at the local level. The MDGs failed to meet most of their planned targets, and have now been succeeded by the Sustainable Development Goals (SDGs). There is a sense that the NUA represents another set of international norms that are worth striving for, yet are undermined by realities in developing countries. The reality in Zimbabwe is that central government—and urban local authorities—have lacked capacity to sustainably manage urban development. Urban councils have not been able to cope with the demand for infrastructure (e.g. roads, water and sewerage) and social services (e.g. schools, clinics and recreational parks). Lost in the quest for universal solutions to particular contexts is any modest reflection that the introduction of international standards into any setting rarely works as imagined The lack of clarity on how to bring sustainable development to sub-Saharan Africa has been ascribed to the intellectual emptiness at the core of the development discourse (Bromley 2008).

Achieving the NUA in Zimbabwe requires the development of governmental capacity for dealing with rapid urbanisation, informality, inequality and increasing poverty in the built environment. Shaping the built environment sciences and practice can substantially contribute to the achievement of the Agenda. A shift towards technological innovation has the potential to make the NUA a reality. Technology offers a bridge between the lagged and restrictive underlying conditions existing in urban centres towards a more inclusive and sustainable built environment. It will change the ways energy, water and transport systems function towards more sustainability. The current multidisciplinary approach to education requires a paradigm shift towards interdisciplinarity in order to give more impetus to the NUA. The introduction of mechanisms for coordination and collaboration among built environment institutes is also at the core of needed changes.

However, the success of technological innovation in the built environment will depend on coherent policies that are founded on political stability: effective government, regulatory quality, rule of law and control of corruption. Corruption is a running theme in the book chapters and its control would undoubtedly set the country on a more vibrant trajectory towards the NUA. Across all sectors of the economy, legal reform is long overdue and should provide incentives for investment in innovation.

Nevertheless, the Covid-19 global pandemic has sparked a rethinking in the way that the built environment is thought about and fostered. This work has grappled with the question of what the built environment could look like considering the possibility of infectious diseases. While the NUA emphasises densification of development, evidence shows that Covid-19 spreads more in densely built environments. These novel issues provide challenges that require new models of urban design. Moreover, the Covid-19 pandemic also forced the world to experiment with working from home. This has huge implications on how places of work and living are thought about.

References

Bibri SE (2020) The eco-city and its core environmental dimension of sustainability: green energy technologies and their integration with data-driven smart solutions. Energy Inf 3(1):1–26

Bromley DW (2008) Formalizing property relations in the developing world: the wrong prescription for the wrong malady. Land Use Policy 26:20–27

Cairns G (2008) Advocating an ambivalent approach to theorizing the built environment. Build Res Inf 36(3):280–289

Chavunduka C (2020) The use of land value capture instruments for financing urban infrastructure in Zimbabwe. In: De Vries WT, Bugri J, Mandhu F (eds) Responsible and smart land management interventions. CRC Press

Chavunduka C, Chaonwa-Gaza M (2021) The political economy of urban informal settlements in Zimbabwe. In: Matamanda AR, Nel V, Chirisa I (eds) Urban geography in postcolonial Zimbabwe: paradigms and perspectives for sustainable urban planning and governance, pp 287–305

Hanson KT, Kararach G, Shaw TM (2012) Rethinking development challenges for public policy: insights from contemporary Africa. Palgrave Macmillan

Iyer-Raniga U, Huovila P (2020) Global state of play for circular built environment. A report on the state of play on circularity in the built environment across Africa, Asia, Europe, Gulf Cooperation Council countries, Latin America and the Caribbean, North America and Oceania. Final report October 2020, United Nations One Planet Network Sustainable Buildings and Construction Programme

Kitchener RF (1999) The conduct of inquiry: an introduction to logic and scientific method. University Press of America, Oxford

Kiviniemi A (2009) Building information models: future roadmap. In: Newton P, Hampson K, Drogemuller R (eds) Technology, design and process innovation in the built environment. Spon Press, pp 141–157

Koskela L (2008) Is a theory of the built environment needed? Build Res Inf 36(3):211–215

Leicht A, Heiss J, Byun WJ (eds) (2018) Issues and trends in education for sustainable development. UNESCO Publishing, Paris

MacLennan M, Stacey J (2016) A new urban paradigm: pathways to sustainable development, Policy in Focus, No. 37

Newton P, Hampson KD, Drogemuller R (eds) (2009) Technology, design and process innovation in the built environment. Spon Press, London

Poncian J, Mgaya ES (2015) Africa's leadership challenges in the 21st century: what can leaders learn from Africa's pre-colonial leadership and governance? Int J Soc Sci Stud 3(3):106–115

Roxburgh C, Lund S, Piotrowski J (2011) Urban world: mapping the economic power of cities. McKinsey Global Institute, New York, NY

Sarkissian W (2005) Stories in a Park: giving voice to the voiceless in Eagleby, Australia. Plan Theory Pract 6(1):103–117

UN-Habitat (2009) Global report on human settlements 2009: planning sustainable cities. Earthscan, London

Unya IU, Akputa GO, Adaobi N (2022) Poverty and leadership failure in Africa: an appraisal. Zamfara J Polit Dev 3(1):1–10

Watson V (2009) The planned city sweeps the poor away…: urban planning and 21st century urbanisation. Prog Plan 72:151–193

SPRINGER NATURE

GPSR Compliance

The European Union's (EU) General Product Safety Regulation (GPSR) is a set of rules that requires consumer products to be safe and our obligations to ensure this.

If you have any concerns about our products, you can contact us on ProductSafety@springernature.com

In case Publisher is established outside the EU, the EU authorized representative is:

Springer Nature Customer Service Center GmbH
Europaplatz 3
69115 Heidelberg, Germany

The manufacturer's authorised representative in the EU is Springer Nature Customer Service Centre GmbH, Europaplatz 3, 69115 Heidelberg, Germany. If you have any concerns regarding our products, please contact ProductSafety@springernature.com

Printed and bound by CPI Group (UK) Ltd, Croydon, CR0 4YY
01/10/2026
02225862-0003